Technology, Society and Inequality

Steve Jones
General Editor

Vol. 87

The Digital Formations series is part of the Peter Lang Media and Communication list.
Every volume is peer reviewed and meets
the highest quality standards for content and production.

PETER LANG
New York • Washington, D.C./Baltimore • Bern
Frankfurt • Berlin • Brussels • Vienna • Oxford

Technology, Society and Inequality

NEW HORIZONS AND CONTESTED FUTURES

Erika Cudworth,
Peter Senker *and*
Kathy Walker, *Editors*

PETER LANG
New York • Washington, D.C./Baltimore • Bern
Frankfurt • Berlin • Brussels • Vienna • Oxford

Library of Congress Cataloging-in-Publication Data

Technology, society and inequality: new horizons and contested futures /
edited by Erika Cudworth, Peter Senker, Kathy Walker.
pages cm. — (Digital formations; vol. 87)
Includes bibliographical references and index.
1. Information technology—Economic aspects.
2. Information technology—Social aspects.
3. Capitalism—Moral and ethical aspects.
4. Technological innovations—Social aspects. 5. Equality. 6. Digital divide.
I. Cudworth, Erika. II. Senker, Peter. III. Walker, Kathy.
HC79.I55T439   303.48'3—dc23   2013003323
ISBN 978-1-4331-1971-2 (hardcover)
ISBN 978-1-4331-1970-5 (paperback)
ISBN 978-1-4539-1042-9 (e-book)
ISSN 1526-3169

Bibliographic information published by **Die Deutsche Nationalbibliothek**.
Die Deutsche Nationalbibliothek lists this publication in the "Deutsche
Nationalbibliografie"; detailed bibliographic data is available
on the Internet at http://dnb.d-nb.de/.

This book is dedicated to the memory of Professor Christopher Freeman who inspired people throughout the world to think about technology and inequality

CONTENTS

Acknowledgments

The authors and editors would like to thank both Mary Savigar at Peter Lang, and the series editor, Steve Jones, for their interest in, and support with, this project. The editors would also like to thank Jacky Senker for all her help in spotting missing references and compiling the references.

Glossary

3G	Third generation
4G	Fourth generation
AGA	AngloGold Ashanti
BBC	British Broadcasting Corporation
BECTU	Broadcasting, Entertainment, Cinematograph and Theatre Union
BWA	Broadband Wireless Access
CCTV	Closed Circuit Television
CIWF	Compassion in World Farming
CMC	Computer-Mediated Communication
CEO	Chief Executive Officer
COTS	Commercial off-the shelf software
CRM	Customer Relationship Management
CT	Computerized Tomography
DESD	(UN) Decade of Education for Sustainable Development
DTT	Digital Terrestrial Television
EEI	Egyptian Education Initiative
ESIB	The National Union of Students in Europe
ERP	Enterprise Resource Planning
EU	European Union
FCC	Federal Communications Commission
FPS	Fair Price Shops (India)
G8	Group of Eight: an international forum for the governments of Canada, France, Germany, Italy, Japan, Russia, the United Kingdom and the United States
GATT	General Agreement on Tariffs and Trades
GDP	Gross Domestic Product
GNI	Gross National Income
HAART	Highly Active Anti-Retroviral Treatment (of HIV/AIDS)
HACAN	Heathrow Association for the Control of Aircraft Noise
HDTV	High Definition Television
HEI	Higher Education Institution
HIV/AIDS	Human Immunodeficiency Virus/Acquired Immunodeficiency Syndrome
ICT	Information and Communications Technology

IFC	International Finance Corporation
ILR	Independent Local Radio
IMF	International Monetary Fund
IRS	US Internal Revenue Service
IT	Information Technology
ITU	International Telecommunications Union
LDC	Least developed country
LOs	Learning Objects
MAFF	Ministry of Agriculture Fishing and Farming (UK)
MIGA	Multilateral Investment Guarantee Agency
MIGs	Manufactured Intermediate Goods
MMS	Multimedia Messaging Service
MNO	Mobile Network Operator
NAFTA	North American Free Trade Agreement
NGO	Non-Governmental Organization
NHS	National Health Service (UK)
NRA	National Regulatory Authority
Ofcom	Office of Communications (UK communications regulator)
OECD	Organisation for Economic Co-Operation and Development
ONS	Office of National Statistics (UK)
PC	Personal Computer
PEPFAR	President's Emergency Plan for AIDS Relief (USA)
PDS	Public Distribution System (India)
PETA	People for the Ethical Treatment of Animals
PMSE	Programme-Making and Special Events
PPP	Public–Private Partnership
PR	Public Relations
SAIL	Stanford Artificial Intelligence Laboratory
SMS	Short Message Service
SSN	Smart Schools Network
SoP	Systems of Provisions
TB	Tuberculosis
UHF	Ultra High Frequency
UN	United Nations
UNAIDS	Joint United Nations Programme on HIV/AIDS
UNESCO	United Nations Educational Scientific and Cultural Organisation
UNIDO	United Nations Industrial Development Organisation
USAID	United States Agency for International Development
VLV	Voice of the Listener and Viewer
WEF	World Economic Forum
WHO	World Health Organisation
WTO	World Trade Organisation

ONE

Introduction: Contested Futures

Technology, Inequality and Progress

Erika Cudworth, Peter Senker and Kathy Walker

The future is contested and there are fierce debates as to where 'we' might be going and how we might get there. According to the World Bank (2006), people living in the United States have 40 percent more income than those living in European Union countries, but they live, on average, two years less. The goals of individual economic wealth (money) and the outcomes for human well-being (indicated by longevity) are very different and involve alternative understandings of what constitutes 'progress.' Important in such debates are judgments about technological, economic and social developments. In turn, these are shaped by ideas about the kinds of social arrangements in which technologies emerge and about how progress occurs.

The most common understanding of progress is in terms of economic development and economic growth. Economic development is expected to increase average incomes, and thereby, standards of living. This is often seen as a neutral goal, and it underpins the policies of nation states and global institutions of financial governance (Stiglitz, 2002). At the centre of this book, however, lies the perception that the primary purpose of capitalist production and distribution is not to satisfy human needs but to create profit for the owners of capital. The fundamentals of capitalist ideology and the predominant strands of economic theory have remained broadly consistent with those articulated by Adam Smith more than three hundred years ago. These can be summarized in terms of three beliefs: first, in the benefits of economic growth for all humankind; second, in the benefits of the application of the principles of the division of labour; third, in the potential benefits of competition among numerous suppliers to meet the needs of even more numerous consumers. These three assumptions fail entirely to reflect the realities of modern capitalism, and huge inequalities characterize contemporary society.

There are, however, other important ways in which progress is understood. These include human rights agendas, human well-being and social equality. The understanding of progress in terms of social equality, and concomitantly, a critique of persistent forms of social inequality, frames the contributions for this book. Forms of social inequality are linked to various social differences, not only those of income and wealth. These include gender, sexual orientation, able-bodiedness, 'race' and ethnicity, religious identity, linguistic community and age. The multiple forms of social inequality vary in intensity and form depending on local, national and global context. Both the interrogation of capitalist production and distribution, and the critique of multiple social inequalities, can be seen as part of a particular political project—of socialism (broadly defined).

This book focuses on three important sets of changes that have taken place in the last fifty years. First, we have seen the increased global spread of capitalist relations to a point at which it has become the dominant form of economic organisation in the world. Second, we have seen the growing significance of rapid technological change in relation to all aspects of economic activity. Third, we have witnessed the concentration of economic power and activity into the hands of an increasingly small number of huge corporations spanning the world. Earnings of poorer workers have fallen relative to the best paid in some of the richest countries in the world, such as the United States (US) and United Kingdom (UK), where poor workers live increasingly vulnerable lives (Ehrenreich, 2001; Toynbee, 2003). Polarities of wealth and poverty have sharpened and South Africa and Brazil are amongst the 'most unequal' societies in the world (Cobbett, 2011). Economic growth has not satisfied the needs of the majority in terms of standards of living and quality of life. The concept of freedom is central to the ideology of democratic capitalism, but it is restricted to civil liberties, consumption and leisure. The alienation, subjugation and poor working conditions experienced by most of the world's workers as consequences of their roles in production are useful to capital, as this leads people to seek to achieve freedom and fulfillment through consumption, which they cannot find in work. In addition to class inequalities and differences of wealth, there are other cross-cutting inequalities associated with region and nation, age and generation and many more, which undermine assumptions of social and economic 'progress' (Walby, 2009). Clearly, inequalities are, to some extent, politically constituted, with government policy and regulation playing key roles in either addressing or exacerbating structures of inequality. The growth of corporate influence on the deregulation of key economic sectors is evidence of the inequality of the policy-making process and the advantages secured by those groups more favourably located in the market economy.

Capitalism has generated significant and enduring structures and techniques that have produced and sustained inequality both within nation states and regions as well as among them. Both the contexts and the unpredictable, often unintended trajectories of global capitalism include major threats, such as very serious environmental problems. Increasing violence (both armed and symbolic), as well as persistent poverty, characterize our present and future horizons. The reemergence of recession in the G8 economies, coupled with the ebb and flow of the fortunes of the world's various regions, have served to remind us of the nature of the global economy that has arisen, which we argue to be

- Inherently unsustainable

- Excessively unpredictable

- Inequitable in relation to the various groups into which humankind is split: among regions and nation states and among social classes within nation states

- Damaging to the environment and destructive of biodiversity.

Capitalism and Growth

Economic growth is measured in terms of GDP, which is predicated on the transfer of increasing proportions of human activity from the realms of subsistence and self-reliance to those of commodity production and acquisition through market mechanisms. These processes require that human need be satisfied increasingly through market-mediated consumption. The development of the particular kinds of intricate and sophisticated techniques of production, distribution and consumption has helped shape the structured inequalities that have emerged. At the same time, processes are occurring that lift significant proportions of the world's population out of subsistence poverty and improve their life chances, including life expectancy. This mixed bag of fortune is one of the classic characteristics of capitalism, as identified by Marx, among others. Although contexts have changed dramatically, it is interesting to note how useful some of the thinking that came out of European nineteenth-century industrialism, and the modern social theory of the twentieth century, remain today.

The so-called 'credit crunch' lays bare the association between modern capitalism and the drive to fetishised consumption, with its mixture of exploitation and desire. There is a general desire for economic growth even in (or perhaps especially in) countries in which, on average, people are already prosperous. In these countries, the majority of the electorate is now relatively prosperous, but there is also a relatively small underclass that is excluded from the benefits of the consumer society by poverty and unemployment. Accordingly, democratically elected governments fear an increase in direct taxation on the relatively prosperous, in the belief that such measures would lead, inevitably, to those politicians losing the next election. Governments of advanced countries place the main burden of paying off countries' debts on the poor and unemployed, and this is a significant factor in causing violent demonstrations and riots. At the same time, a shortage of investment capital generally presents a severe challenge to developing economies. An exception to this is China, the present 'workshop of the world,' where a huge working population is exploited to generate enormous surpluses, a substantial proportion of which are invested abroad, rather than being invested at home to improve the lives of the Chinese people.

Even before banks in advanced countries expended billions of dollars in subprime loans, which poor borrowers were never going to be able to service, a consumer boom in 'advanced' nations, fueled by loans from China, was widely viewed as positive in some senses, but essentially unsustainable. The false boom expanded markets for the products of the manufacturing hub and for services from everywhere, but the imbalance between production and consumption areas could not last. The availability of finance through the increased range of financial instruments developed by the banking and securities industry stimulated and accelerated a process that was already in progress.

Fundamentally, modern capitalism is predicated on both increasing consumption and the persistence of inequality (in the context of overall scarcity). The two are intimately related. The stimulation of rapid growth in consumption requires that people have 'aspirations' to possess goods and services that they do not have at present. For this to continue indefinitely, some people must have more than others in order to demonstrate what might be aspired to and to achieve social identities that differentiate among social groups. To be sure, there are drivers of consumption other than aspirations and identity—such as genuinely new and better products, improved techniques, cost savings and other indicators of substantive innovation as well as on a more micro level, the local economic development that contributes to social and community cohesion. But one of the dominant groups of techniques of the production and consumption that has fuelled the boom of the last twenty years has been based around differentiation and market segmentation and hence related to the inequality-aspiration-consumption cycle. It may, of course, be possible to have differentiated consumption that is based on real differences—in taste, environment, and so on. However, the fetishisation of style, choice and individuality goes beyond this.

Under the current capitalist system, innovation is much faster than it might otherwise be, and the lifespan of the commodities it produces is more temporary and transient in nature. What has emerged, therefore, are its design faults of waste and its tendency toward rapacious behaviour in terms of the stresses it places on the environment. The biggest problem is that it has become self-sustaining at the level of culture. Despite the current crisis, aspirant people, regions and nations are unlikely to stop wanting to emulate the lifestyles of the wealthier; given the theory of relative poverty and the drive to improve life chances of 'me and mine,' why should they? The authors of this volume enjoy the advantages of our intensive education and being well-fed, shod and clothed, as do many of our readers.

Persuading the wealthy to consume less voluntarily is not going to get onto the agenda of many governments any time soon, even in the light of compelling evidence that living in very unequal societies, such as the US, does not even benefit their wealthiest inhabitants (Wilkinson & Pickett, 2009). Promoting the reduction of consumption is outside the current paradigm, not least when the understanding of the economic model that fuelled the postindustrial boom was precisely that consumption was the key economic driver—the path to salvation for exhausted industrial economies that had relied on the poverty of the working class and the colonial system in order to produce a surplus. The unfortunate and potentially calamitous issue is that the consumption boom was not driven by citizens spending their savings but by citizens spending the credits they were advanced, financed by the free and easy flow of apparently never-ending cash.

This flow of cash has an interesting progeny—amassed by a curious mixture of financial instruments (instrument shall rely on instrument and hopefully never get called in at the wrong moment), and curiously old-fashioned occupations such as the arms and drugs trade (albeit reinvented as just wars fought offshore from Europe and the surveillance and security industries). Also in the mix is the brutal exploitation of people living in subsistence conditions and therefore vulnerable to sweatshop employment, or the offspring of generations who experienced such, and who were therefore driven into the upgraded but somewhat intensive if not well-paid labour of proper jobs in proper factories and clean jobs in call centers.

Economic growth and ever-increasing consumption are fundamental to modern capitalism. The questions asked in this book are whether the current models of economic growth, with their requirement for the continual replacement of collective consumption and subsistence production by the individualization of consumption, can promote the reduction of inequality or the longevity of social life on a sustainable basis. It is also increasingly apparent that substantial reduction of man-made environmental degradation is unlikely to be achieved if the economic system is based fundamentally on ever-increasing consumption.

While economic growth has, indeed, brought substantial benefits in terms of improvement of the standards of living of billions of people, this book questions the general assumption that its continuation can result in the elimination of the large continuing reservoirs of dire poverty. Our attempt to supplement some profound recent analyses of such issues (Collier, 2008; Sachs, 2005) is based on the belief that some further substantial changes in policy direction are required to prevent the persistence and recurrence of large populations living in poverty. The drive to economic growth has been a dominant policy discourse at various levels of polity—local, national and global. It represents a particular version of what 'progress' might be and is currently articulated through the specific articulation of capitalist relations that is neoliberalism (Harvey, 2005). Herein, the notion of the effectiveness of the market is elevated as the mechanism for achieving progress, and the freeing of the market from state controls is understood as the most expedient way to ensure economic growth and thereby, enhance human well-being and freedom (England & Ward, 2007; Hayek, 1960). However, this record of neoliberalism on achieving economic growth is contested (Stiglitz, 2002), and there are other, competing, comprehensive visions of what might take us toward a 'better' future. The project of social democracy, for example, articulates the relationship between the achievement of 'progress' and the need to lessen forms of inequality (see Giddens, 2001) and realise human rights (Held, 2004).

Inequality, Technology and Progress

According to the OECD's 2008 report, *Growing Unequal? Income Distribution and Poverty in OECD Countries*, the gap between rich and poor has grown in more than three-quarters of OECD countries over the past twenty years, and the decades of economic growth before the current recession have disproportionately benefited the rich more than the poor. This is supported, for example, by research from the Institute for Fiscal Studies in the UK, which reported in 2009 that income inequality in the country had "risen (on most measures) in each of the last three years and was now at its highest level since our comparable time series began in 1961" (Brewer, Muriel, Phillips, & Societal, 2009). Research shows that economic inequality, that is, disparities in the distribution of economic assets and income, is closely related to social inequality and an individual's access to social goods (housing, education and healthcare) and social standing. A US Internal Revenue Service (IRS) report issued in October 2007 revealed record levels of social inequality in the US and a growing level of polarisation between rich and poor. The report, based on tax returns from 2005, revealed that America's wealthiest one percent of the population accounted for twenty-one percent of all income, whilst the bottom fifty percent earned just 12.8 percent (Van Auken, 2007).

These reports reveal unprecedented levels of inequality, not only between rich and poor nations, but also between rich and poor citizens within some of the wealthiest countries in the world. Economic development and material living standards are very important for people living in developing countries, especially for the poor and the poorest. The early stages of economic growth make people happier, but happiness fails to increase after a certain level of income per head has been reached. For example, as countries start to get richer, the number of people afflicted by infectious diseases declines. Within individual countries, richer people do better in terms of life expectancy, health and social problems than poorer people. But, as already rich countries become richer, "diseases of affluence," such as heart disease, stroke and obesity, start by afflicting the rich, but then go on to afflict the poor more than the rich. There is evidence that "some of the most affluent societies seem to be social failures" (Wilkinson, 2008).

For such reasons, Wilkinson and Pickett (2009) suggest that people in most developed countries can't get much more benefit out of economic growth. Measures of well-being and happiness in these countries have ceased to increase. Some have argued that this requires a reorientation of priorities away from growth to well-being or 'happiness' (Layard, 2005). Indeed, the drive to increase incentives to work may actually lead to increased inequalities that compromise human well-being (Wilkinson, 2005). Moreover, there have been long-term rises in numerous social problems, such as violence and crime, in rich countries. While life expectancy continues to increase in developed countries, such increases do not have any direct relationship with economic growth. For example, the US is far richer than Greece, but average life expectancy is no greater. Sooner or later, as developed countries become richer, countries reach a level of average income at which 'diminishing returns' set in and additional income yields fewer and fewer additional benefits in terms of health, happiness or well-being. While there is no direct relationship between the welfare of the people and societies in developed countries, there appears to be quite a close negative relationship between the welfare of societies and the extent of income inequality. In developed countries, where there is a high degree of income inequality, individuals tend to experience poor health and mental health, a shorter life expectancy, a higher proportion of imprisonment, obesity and lower literacy scores (Wilson & Pickett, 2009).

Indeed, concerns about rising levels of inequality have resonated with governments in OECD countries, both because of their wider implications for social development and cohesion and because they have occurred in many of them, despite considerable increases in government spending on social benefits to offset the growing income disparity. Ed Miliband, currently leader of the opposition Labour Party in the UK, has stressed the importance of addressing growing social inequality, arguing that

> poverty, social mobility, and the ability to lead a freely-chosen life—are all in a sense aspects of an egalitarian position, a position relating to equality. Not just income poverty, but social mobility, which is about the unequal distribution of opportunity, and the opportunity to lead the life people want to lead, which is about the unequal distribution of freedom. (Miliband, 2008, p. 3)

The OECD report *Growing Unequal? Income Distribution and Poverty in OECD Countries* concludes that the effectiveness of income redistribution via tax and benefit systems has largely declined in the last ten years and places emphasis on increasing employment and improving education as the most powerful ways to boost income and eradicate inequality. Whilst the call for improved educational opportunities cannot be questioned, the objective of increasing employment and job creation in a recession-hit, global economy may prove a more challenging recommendation. In addition, the ineffectiveness of government tax and social policies in addressing income and social inequalities, and the fact that these fundamental problems are occurring in otherwise largely (until recently) successful Western economies, call into question the ability of modern capitalist market economies to support equal and fair social structures.

The achievement of 'progress,' in terms of the promotion of human as well as economic development, the avoidance of excessive inequality and the provision of services that enhance well-being (such as education, health and [other] care) in order to secure social justice, is the aspiration of social democracy, broadly defined. Even a 'social democratic' future and policy project is contested. For example, the rearticulation of social democracy as the 'third way' of the 'new' Labour Party in the UK has been suggested as an appropriately contemporary form of social democracy (Crouch, 1999; Giddens, 1998) or as an effectively neoliberal development in its reliance on the market as a means of social allocation (Arestis & Sawyer, 2005; Jessop, 2002). Certainly, the rearticulating of 'inequality' as 'social exclusion' was a dilution of social democratic principles (Lister, 1998), and this can also be seen as a move away from a concern with the profound inequalities in the distribution of wealth in capitalism. In this book, we are very much concerned with such inequalities, and we aim to address issues of inequality that affect both developed and developing countries, not just within the framework of income distribution but, crucially, by exploring the wider dynamics of capitalist systems of production and consumption.

This book has a specific focus, however, because we consider the established paths and future developments in terms of the relationship among inequality, technology and society. It develops and extends the contribution we have made to this small but growing literature, in two previous books written by different but overlapping groups of authors. In *Technology and In/Equality: Questioning the Information Society* (Wyatt, Henwood, Miller, & Senker, 2000), the authors explored the social, cultural, economic and political contexts that shape and are shaped by technological developments, challenging some of the fundamental ideas behind the concept of the information society and the uncritical analysis of the widespread and indisputable benefits associated with it. In *The Myths of Technology: Innovation and Inequality* (Burnett, Senker, & Walker, 2009), the authors explored ways in which mythic ideas about technologies underpin our understanding and expectations of them and often lead to unfulfilled potentials and greater social inequality when commercial market pressures are the primary motivations for their development. The focus of this book continues the trajectory of our earlier research by concentrating on the ways in which technology has the potential to alleviate many of the features of poverty and inequality but is frequently developed in such a way as to exacerbate the social inequalities that already exist and, in the case of some technologies, give rise to completely new types of inequity and unfairness. The chapters in this book draw together some tentative answers to some central questions, which arise from this examination of technology in our technology-dependent global economy. These include the following:

- Is inequality endemic and in some ways an essential component of a successful competitive, commercially-orientated capitalist economy dependent on consumerism and capital accumulation?

- What roles can/does technology play in exacerbating or alleviating inequality?

- Is there a need for specific policies to influence the path of technological development in directions that are more favourable to the poor, principally those living in developing countries but also poorer people in developed countries?

- If so, what is the nature of those policies and what forces could be created toward their development and implementation?

During the last century, the principal forces that have driven the choices of which technologies are developed and the directions in which they are exploited are the requirements of multinational corporations to increase their profits, and the requirement, mainly by governments, to expand their economies, encourage investment and enhance their economic and military strength. Sometimes, as in the case most notably of information and communications technologies, the path of technological development has been strongly influenced by interaction among these motivations. Increasingly, these motivations are combined with the motivation by governments to create incentives for the development and exploitation of technologies that benefit the environment, principally in terms of their perceived potential for curbing global warming. To a very considerable extent, the implications of technology for the distribution of benefits and costs between rich and poor have been largely incidental to these principal motivations. Nevertheless, the international research agenda is now dominated by technocratic and technologically determinist views of innovation. Such views obscure the role that social forces and conflicting interests play in the development of science and technology and in determining how science and technology interact with society.

In recent years, concepts such as those of the 'Information Society,' the 'New, Digital or Knowledge Economy' and 'Globalisation' have transcended social science and have been adopted as incontrovertible paradigms by the majority of policy-makers and by international organisations such as the World Bank, the World

Trade Organisation, the European Commission, the OECD and the G8. These paradigms have been used to justify the inevitability of the policies being pursued (de Miranda, 2009). They also dominate academic agendas worldwide—increasingly they require universities and other educational and research institutions to demonstrate the 'relevance' of their work to the kind of society prescribed by such concepts. The inevitability of the predicted changes is often justified by ascribing to technology the role of social driver, contrary to the principal approach taken in social studies of science and technology. Yet in the same period in which this new society is deemed to have been developing, the social and economic gulf that separates the rich from the poor has greatly increased both within nations and among nations. Large sections of global society feel marginalised and disenfranchised rather than empowered, as those who are optimistic about technology would have us believe.

This book aims to contribute to an alternative research agenda that will continue to challenge dominant paradigms. Its aim is to investigate the causes of growing inequality and the roles that science and technology play in this. We hope to provide the basis for the development of policy proposals designed to counter the inequality and injustice created by the ways in which science and technology are used at present; we attempt to identify some of the political and other barriers to implementation of such policies. In doing this, we echo Wilkinson and Pickett's (2009) conclusion that, "the benefits of greater equality seem to be shared across the vast majority of the population" (p. 186).

Production, Consumption and Choice

Early forms of capitalist mass production were based around and dependent on mass consumption—mass production, driving mass consumption, driving mass production. The limitations of the production line process provided little choice among available products—competition, and therefore the limited amount of choice available for consumers, was provided largely by competitors in the market. Nevertheless, the ability to buy into the consumer society, the acquisition of a car and consumer electronics, was seen as growing evidence of increasing individual and social affluence. The adage, "Any man who finds himself on a bus after the age of 30 can count himself a failure,"* clearly identifies success with consumption: the ability to buy a car, and therefore, the ability to earn enough to buy that car. The notion of inequality is clearly created by the ability of some to choose private as opposed to public transport, via the process of consumption—the haves and the have-nots.

Collective consumption places limits on the growth of markets. It is, therefore, in the interests of capitalism to promote economic growth by individualizing

---

* This adage is often credited to Margaret Thatcher when she was Conservative Prime Minister, but the originator of the quote is more reliably acknowledged as Loeila, Duchess of Westminster.

consumption as much as possible. Promotion of the choice agenda can therefore be understood in this context as the promotion of the individualization of consumption that capitalism requires to survive and prosper as a system. This is the case whether we are referring to the most recent products of 'light capitalism' (Bauman, 2004) such as software, internet services and virtual goods, or the traditional 'heavy' goods, such as cars and home electronics. The growth of markets is also restricted by the persistence of subsistence production. This can be understood to mean more than self-production to meet the basic human needs of food, clothing and shelter but to include also the fulfillment of cultural needs such as music and dancing, painting, fashion, sport and so forth, which in precapitalist social formations and in less advanced forms of capitalism have been traditionally carried out outside the realm of the market. It is this vein of thinking that Amartya Sen (1999) has drawn upon in developing his influential 'capabilities' approach to human development, which stresses human well-being through the fostering of 'capabilities.' This represents a significant challenge to a focus solely on economic growth measured in terms of GDP, but Sen is ambiguous on questions of equality and inequality. He defines capabilities as the 'substantive freedoms' to 'choose a life one has reason to value' (pp. 74–75). Here, as Sen puts it, development is constituted through freedom, but we must also be free to choose outcomes that may be unequal. Sen argues that the emphasis on various paths to progress and on the importance of choice accounts for different historical, geographical and cultural contexts and enables the avoidance of 'inappropriate' development. Whilst this emphasis on noneconomic progress might invite a critique of market-led models, the prioritisation of choice over equality makes it difficult to evaluate outcomes (Walby, 2009, p. 9). In addition, building the kinds of capabilities to make real choices within globalized capitalism is problematic.

At the centre of this book lies the perception that the primary purpose of capitalist production is not to satisfy human needs but to create profit for the owners of capital. In capitalist ideology, labour is not conceptualised as something human beings engage in but as a factor of production. What is to be human is confined to the sphere of leisure and consumption when work has been completed. The concept of freedom is central to the concept of democratic capitalism, but it is restricted to civil liberties, consumption and leisure. The alienation, subjugation and poor working conditions experienced by most of the world's workers as consequences of their roles in production are useful to capital, as this leads people to seek to achieve the freedom and fulfillment they cannot find in work through consumption. Whilst capitalism cannot be said to have saturated all regions and parts of the globe to the same extent, it is an inherently globalizing system of economic production and exchange and of social relations (Giddens, 1990).

Improvements in technology, particularly flexible production and just-in-time technology introduced in the 1980s, have resulted in a far wider range of mass-produced products on the market. In addition, the global division of labour and relocation of production to third-world countries has meant the cost of production has remained low (see examples of Primark and others). 'Choice' has now become the mantra of capitalist market economies. Apparently, we have more choice, more quickly (faster turnaround) and ever more cheaply. Since choice based on cost automatically disadvantages some people, inequality appears unavoidable within this capitalist structure. However, it is difficult to argue that 'choice' in itself is necessarily problematic, since the whole ethos of free, democratic countries is based around just such a notion of free 'choice' in fundamental matters, such as our elected government, to choose or not to choose to be religious, to choose our profession, our way of life, our life partner and so on. 'Choice' has become the imperative of postmodern capitalism. This applies not only to mass-produced commodities, from baked beans to trainers to consumer electronics, but increasingly to cultural artefacts in the realms of television and media more generally. Our current choices help us create our identities and underpin our individuality seen particularly in the relationship between youth identity and its construction through clothes, fashion and music.

Yet despite the apparent plethora of choice in the marketplace, when choosing clothes, consumer electronics, food and even cultural artefacts such as music and film, the reality is that there is probably less choice of different varieties of these products. George Ritzer (2004) has argued this lack of choice assumes increasingly globalized characteristics as societies around the world move from the consumption of 'something,' that is, goods and services that are rich in distinctive content, to the consumption of 'nothing,' goods and services that are relatively devoid of distinctive content. This form of consumption is enabled and promoted, he suggests, by the growth and international expansion of huge multinational corporations. Most of the so-called choices made available to us are the products of such corporations, which have introduced market segmentation (as in the case of clothing manufacturers) or niche marketing or channels (in the case of television and music distribution) to replace real competition and choice. For example, this would include Time Warner and News Corporation in the media and audiovisual fields and Arcadia Group Ltd (Burtons, Dorothy Perkins, Evans, Miss Selfridge, Topshop, Topman and Wallis) in the retail clothing sector. For several years in the television and broadcasting sector, commercial companies have lobbied relentlessly for the opening up of formerly public service broadcasting sectors to market competition. These arguments were generally on the grounds of increased choice and consumer sovereignty. Ironically, these calls for the breakup of public service sectors were generally taking place at the same time as even greater consolidation in the commercial domain. These perceptions are supported in different ways by the various chapters in the book.

The Structure of the Book

In this book, we address issues of inequality that affect both developed and developing countries. In doing so, we are not just concerned with the framework of income distribution but explore the wider dynamics of capitalist systems of production and consumption. We examine the dimensions of inequality from both an economic and sociocultural perspective. The book comprises three sections. The first considers the relations between technologies, inequalities and exploitation, and the second focuses on issues surrounding technologies and developments. The final section considers the contested nature of technologies and associated opportunities, in the face of the future.

Part I examines the alienation, low pay, subjugation and poor working conditions experienced by large groups of people and also by nonhuman animals in industrial agriculture. It considers how regimes of inequality continue to be reproduced in production systems and through the supply chains of goods. These goods—the personal computer, the mobile phone and meat- and animal-derived foodstuffs—are often presented as necessary, enabling and desired products. The technics of their production and the nature of their use, however, tell a rather different story. In this section, Chapters 2, 3 and 4 consider the systems of production, distribution and choice in relation to the delivery of the commodities of food, mobile communications and computing. Technological change has been rapid throughout these sectors, but we show that humankind in general would benefit from radical rethinking of the direction of change. The section also examines the alienation, subjugation and poor working conditions experienced by most of the world's workers as consequences of their roles in the realm of production. This is potentially useful to capital as it leads workers to seek to achieve the freedom and fulfillment that they cannot find in work through consumption.

In Chapter 2, Alvaro de Miranda shows us the ways in which exploitation can operate quite subtly, and revealing the problematic inequalities embedded in the production and consumption of personal computers is the subject matter of the chapter. Here, we are shown the disjuncture between the vision of PC pioneers and the production, distribution and consumption of these devices. In Chapter 3, we focus on the supply chains involved in the production of less essential commodities such as jewellery and ICT devices. Richard Sharpe argues that supply chains are not apparent to consumers and the price that, for example, manufacturers of jewellery and electronic products pay for precious metals such as gold do not reflect the huge human costs paid by the miners—men, women and children—who work in terrible hazardous conditions for low pay to mine those metals. In Chapter 4, Erika Cudworth examines the drive to produce cheap meat and animal-based food products as a means of reducing malnutrition and hunger first in the US and Europe since the Second World War, and promoted internationally by the United Nations, which, in the 1960s and 1970s, emphasised the necessity of increasing production and making such food more available in poor countries. These Western government initiatives were driven by the corporate interests of the multinational corporations based in their territories. These initiatives had considerable implications for vulnerable groups of humans exploited as labourers, the climate, and domesticated animals, whose bodies and labour are exploited for food.

In the second part of the book, we focus on issues concerning technology, development and inequality. This includes relationships between ICT use and development, which are considered in two chapters. Education, health services and food production, distribution and consumption play critical roles in development, so this section also includes chapters on these subjects. International organisations such as the World Trade Organisation and the World Bank try to persuade developing countries to invest heavily in ICTs and to open up their public services to international capital. This part of the book serves to cast serious doubt on whether such international organisations' policies actually benefit developing countries.

In Chapter 5, Miriam Mukasa considers the cultural implications of the consumption of ICTs for development. Organisations such as the World Bank try to persuade developing countries that they should invest heavily in ICTs, as they are engines of growth. Such propositions rely on the belief that ICTs can help developing countries narrow the gaps in productivity and output that separate them from industrialised countries—and even that they can 'leapfrog' stages of development into the information economy. But most developing countries are unable to cope with the new technological paradigm or to exploit its potential. Peter Senker considers the social and geographical context of health service provision in Chapter 6, comparing regional differences in resources and evaluating the ability of systems to deliver and deploy healthcare technologies. The problematic assumptions underpinning the use of ICTs for development is also taken up by Allyson Malatesta in Chapter 7, which focuses on the impact of ICTs in education. ICT producers have promoted their products as means through which excellent education can be made available to diverse and widespread communities. The World Trade Organisation believes that public services should be opened up to international capital and that this would benefit both globalization and education. However, critics maintain that further opening up the markets of poor nations to transnational corporations is liable to create greater inequalities between rich and poor nations. In Chapter 8, Peter Senker questions the prevailing vision of a 'modern' agriculture from the Green Revolution to the current Gene Revolution as a standard, preferred pathway to development. Such a perspective centres on technology, production and growth. Key elements of the modern agri-food 'system' involve a wide array of external expensive inputs such as research and development, fertilizers, seeds and irrigation. Centralized, technology-driven economic growth through sustained innovation and trade is envisaged as providing pathways out of agriculture or a shift of subsistence-oriented 'old' agriculture to a modern, commercial, 'new' form of agriculture.

In the third section, authors explore issues relating to the opportunities offered by new technologies and the often very contradictory nature of their development and the uses to which they are put. It is widely recognised that technologies are a catalyst for change, but their potentials can be shaped in many ways and by a rich mix of social, cultural, political and economic factors. In some instances, the radical potentials of technologies may be utilised proactively to challenge existing structures of inequality. This may be the case particularly in the early days of technological development, when technologies may be appropriated in ways that are entirely unpredictable and even undesirable to their original producers. In others, the economic and political processes that more often determine the nature of their development limit these radical potentials and effectively constrain our contribution to, and control over, their development and impacts on our lives.

In Chapter 9, Kathy Walker examines how extensive development in wireless technologies and applications has resulted in a growing demand for deregulation and liberalisation of the radio spectrum to allow wider and more easily accessible access for new businesses and commercial initiatives. ICTs have also been introduced into educational institutions for motivations other than improving teaching and learning. Recent years have seen a rapid expansion of new surveillance technologies, not just in the wider community but also in schools, particularly secondary schools. These include extensive closed-circuit TV (CCTV) monitoring, screening technologies, tracking devices and biometric tools. In Chapter 10, Charlotte Chadderton considers the implications of school surveillance for notions of citizenship, social justice and participatory democracy, using the UK as a case study. In Chapter 11, Maxine Newlands explores how technological advances in communication networks open up new platforms for democratic debates. Technological advances in communication blur the boundaries between traditional journalists and citizen journalists. This raises questions about the nature of political discourse and social movements and the practice of democracy. Drawing on a case study of radical environmental activism movements in the UK, this chapter shows how new technologies have enabled activists to bypass traditional media practices. New social movements made from environmental activists' collectives can produce their own websites and news reports and adapt old tactics through new technology. The result is a narrowing of the inequalities in the production of 'news.' Smart phones, tablets and the World Wide Web provide a new platform for a wider range of voices, which were once limited by hierarchical media practices. The web and new media technologies, as with any new tools of communication, can only decrease inequalities if correctly applied. This chapter therefore considers both the positive and negative implications of new forms of communication. The internet is not a saviour but just one means of changing political and cultural discourse.

The chapters in this book lead us to the perception that the primary purpose of current production and distribution is not to satisfy human needs but to create profit for the owners of capital, and the conclusion, Chapter 12, elaborates on the system's extensive failures in terms of directing technological change in appropriate directions.

PART I

Technology, Inequality and Exploitation

This section of the book examines the alienation, low pay, subjugation and poor working conditions experienced by large groups of people and also by non-human animals in industrial agriculture. It considers how regimes of inequality continue to be reproduced in production systems and through the supply chains of goods. These goods—the personal computer, the mobile phone and meat- and animal-derived foodstuffs—are often presented as necessary, enabling and desired products. The technics of their production and the nature of their use, however, tell a rather different story. In this section, Chapters 2, 3 and 4 consider the systems of production, distribution and choice in relation to the delivery of the commodities of food, mobile communications and computing. Technological change has been rapid throughout these sectors, but we show that humankind in general would benefit from radical rethinking of the direction of change. The section also examines the alienation, subjugation and poor working conditions experienced by most of the world's workers as consequences of their roles in the realm of production. This is potentially useful to capital as it leads workers to seek to achieve the freedom and fulfillment that they cannot find in work through consumption.

In Chapter 2, Alvaro de Miranda outlines the idealistic visions of some PC pioneers who perceived personal computers as agents of liberation for workers from the drudgery of submission to mainframe systems. PCs would meet workers' human needs to express their creativity by using a tool that was under their own control. However, as the industry developed, massive inequalities became embedded in the production and consumption of PCs.

Whilst de Miranda provides us with an insight into the subtleties of inequality, the deep-seated inequalities and forms of exploitation embedded in technologies is the crux of Chapter 3. Here, Richard Sharpe points out that the use of ICTs can be highly beneficial and liberating, allowing millions of people access to ever-widening ranges of information and services and communication with huge and ever-growing numbers of other people. But his chapter is mainly concerned with the dark underside that is an inevitable concomitant of these benefits. Gold is just one of the rare minerals essential for the manufacture of ICT devices. The chapter outlines the huge human costs paid by miners of such materials—men, women and children—who work for low pay in hazardous and unhealthy conditions to mine rare minerals that have to be incorporated in the components of ICTs to enable them to function. The tiny proportion of the price paid for gold by manufacturers that is received by miners does not reflect the enormous human costs that afflict them.

In Chapter 4, Erika Cudworth examines the drive to produce cheap meat and animal-based food products as a means of reducing malnutrition and hunger first in the US and Europe since the Second World War and promoted internationally by the United Nations, which, in the 1960s and 1970s, emphasised the necessity of increasing production and making such food more available in poor countries. These Western government initiatives were driven by the corporate interests of the multinational corporations based in their territories. This had considerable implications for vulnerable groups of humans exploited as labourers, the climate and domesticated animals, whose bodies and labour are exploited for food.

TWO

Technology and the Individualization of Consumption

The Development of Personal Computing

Alvaro de Miranda

On January 24, 1984, during the third quarter of the Super Bowl game between the Los Angeles Raiders and the Washington Redskins, an advertisement was broadcast that became one of the best-known television ads in the history of advertising. Entitled "1984," the advertisement, directed by Ridley Scott, announced the forthcoming launch of the first Apple Macintosh computer, using a dramatic metaphor based on Orwell's eponymous dystopian novel. It starts with an army of androgynous, shaven-headed people dressed in identical grey tunics marching down a tunnel toward a hall where a giant screen is showing a close-up picture of a bald and bespectacled 'Big Brother' droning on about having created "a garden of pure ideology where each worker will bloom secure from the pests of contradictory and confusing truths," and stating, in imperious tones, "We are one people. With one will. One resolve. One cause . . . . We shall prevail." A blond woman athlete, a hammer thrower, in red shorts and a white vest displaying a barely discernible Apple logo, runs into the hall, pursued by a police squad in full riot gear, and hurls her hammer at the screen, destroying it. A voice then reads aloud the legend that appears on the screen: "On January 24 Apple Computer will introduce Macintosh. And you will see why 1984 won't be like '1984.'"

The metaphor used by the advertisement has several dimensions, all of which had been crucial in informing Apple's development of the Macintosh computer and in the company's subsequent marketing strategy. Arguably, they were also essential to the eventual success of the Macintosh computer. One facet of the metaphor is the reference to the suppression of individuality by the power of uniform collective identities. The 'collective will' is enforced through violent repression by a 'Big Brother,' an implicit incarnation of the State, claiming to act on behalf of the collective, the 'we' who 'will prevail.' However, whilst Orwell's *1984* was a satire on the totalitarian collectivist state, Apple wanted the metaphor to be extended to include an attack on IBM, despite the denials of the advertisement's producers. At the annual shareholders' meeting that took place on the same day the ad was broadcast, Steve Jobs, chief executive and co-founder of Apple, stated the following:

> It is now 1984. It appears IBM wants it all. Apple is perceived to be the only hope to offer IBM a run for its money. Dealers initially welcoming IBM with open arms now fear an IBM dominated and controlled future. They are increasingly turning back to Apple as the only force that can ensure their future freedom. IBM wants it all and is aiming its guns on its last obstacle to industry control: Apple. Will Big Blue dominate the entire computer industry? The entire information age? Was George Orwell right? (Brooks, 2006)

In the advertisement, the fight for individual freedom is symbolised by the role of the woman athlete who destroys the image of 'Big Brother.' Apple implicitly aligns itself with this fight and portrays the Macintosh as the means by which individuals will achieve it in the world of computing. The metaphor, in its subliminal identification of IBM with the repressive State of Big Brother, also addresses the business world and, as I shall demonstrate, the world of work. It carries an attack on the role of the big corporation and on monopoly, through its implicit reference to IBM.

However, the implicit promise contained in the advertisement that purchase of an artefact will deliver increased freedom for the consumer in the realm of his or her work conceals an issue which, as I demonstrate in this chapter, is central to the question of inequality. How the artefact itself is produced under a capitalist system so that it is affordable to the consumer raises questions about the relation between the relative benefits to the workers who have produced it and to the consumer who purchased it, as well as to the relative benefits to the worker and the capitalist in its production. As I show, in order to make the artefact accessible to the consumer both in economic and in technical terms, and in order to deliver profits to the capitalist, the wages of the workers producing it have to be squeezed. Their freedom must also be heavily repressed. Thus the benefits that may be realised for some in the realm of consumption are achieved at the cost of the economic well-being and freedom of workers in production.

The promise that the Macintosh would contribute to human freedom was, as I argue, also predicated on its 'user-friendliness,' that is, on the reduction of the technical skill required to use a personal computer. This was one of the major new features it offered. It contributes to the development of another kind of inequality, that between those who possess technical skill and increasingly populate only the realm of the production of the technology and those who thereby become disempowered and for whom technology is nothing but a 'black box' over which they have no control.

The Development of the Macintosh

The Macintosh computer constituted a major conceptual break with the previous generations of microcomputers in relation both to the market it was aimed at and in the way its technology was configured to satisfy the needs of the user in its perceived target market. This break was already being signalled by the 1984 Super Bowl advertisement that addressed its target audience self-consciously through the use of an emotionally arresting metaphor and avoided all mention of the physical attributes of the artefact it was designed to promote.

The nature of the break was that, until the Macintosh was developed, microcomputers were mainly sold to computer hobbyists. The fundamental nature of the hobbyist is that the technological aspects of the artefact constitute a main focus for their interest.[1] A hobbyist enjoys tinkering with the technology as much as, or even more than, using the artefact as a tool for some other purpose. A hobbyist has to have considerable technical knowledge in order to be able to use the artefact. Before the advent of the Macintosh, a considerable amount of technical knowledge was required in order to use a microcomputer.

The Macintosh was specifically designed to be used as an *individual* tool. The essence of a tool[2] is that it should be relatively easy to use. Learning to use a tool should require little diversion of labour time from the primary purpose of that labour. Or, in today's terminology, a tool should be 'user-friendly.'

The individual tool aspect of the Macintosh's design came when designers considered that the hobbyist market was saturated because everyone who wanted a computer already had one (Potter, 1979, p. 60). In order to extend the market, Macintosh had to design a 'user-friendly' tool. To conceptualise the function of the microcomputer that they were trying to design, and to differentiate it from that of a mainframe computer, designers used the metaphor of the relationship between a Volkswagen and a passenger train. The Volkswagen does not go as fast or as far as a passenger train, but with a Volkswagen you can go anywhere you want to when you want to. In this way, they arrived at the conceptualisation of the artefact that they were trying to design as an instrument of individual freedom. This was eventually expressed metaphorically in the Super Bowl advertisement.

The Macintosh marketing campaign that followed the original 1984 advertisement primarily targeted 'knowledge workers,' a concept first introduced by management guru Peter Drucker in 1959. This group was described in an internal Apple memorandum as professionally trained individuals who are paid to process information and ideas into plans, reports, analyses, memos and budgets. They generally sit at desks. They generally do the same generic problem-solving work irrespective of age, industry, company size, or geographic location. Some have limited computer experience—perhaps an introductory programming class in college—but most are computer naive. Their use of a personal computer will not be of the intense eight-hour-per-day-on-the-keyboard variety. Rather they bounce from one activity to another; from meeting to phone call; from memo to budgets; from mail to meeting. Like the telephone, their personal computer must be extremely powerful yet extremely easy to use. *(cited in McKenna, 1991, pp. 191–192)*

The target group of potential customers for the Macintosh computer confirms the designers' conception of the artefact as an individual work tool. The target group includes workers who are self-employed, workers in small organisations, and workers in large organisations. What is common to all these workers is that they all have a considerable amount of autonomy in their work. This would be true even of those who work in large organisations. The Macintosh was intended as an aid to this autonomy, a productivity tool in a labour process that was, to a large extent, self-directed by the worker himself or herself. This freedom to decide the allocation of one's labour time extended beyond the freedom from tight management control for workers in large business organisations and included freedom from dependence on the labour of other workers in achieving the purpose of the labour of the 'knowledge worker.'

The Macintosh never quite achieved its aim of becoming the main tool of all 'knowledge workers.' In the context of the large corporation, it failed to displace the IBM PC[3] and clones. However, it did attain an almost complete monopoly over a subset of this group, the creative worker. For architects, designers, musicians, filmmakers and artists of all kinds, the Macintosh became the computer of choice. This highlights other self-conscious features of the Macintosh's design, which were particularly attractive to creative workers. One such feature is the concealment of virtually all technical aspects of its construction inside a single box. The external case of the Macintosh could not be opened by the user and required a special tool that was provided only to qualified experts. In addition, the smooth box in which all the technology was hidden had been designed to be an attractive object in itself, a demand of Apple's cofounder, Steve Jobs, which has been a constant principle of Apple design to this day. That virtually no technical knowledge was required and the artefact was good-looking was particularly attractive to the creative worker, the antithesis of the technical nerd.

The Macintosh's user-friendliness was mainly achieved through its pioneering use of the graphical user interface.[4] For the first time in a product intended for mass consumption, metaphors of icons and menus were used as simple visual aids that could be pointed to and clicked on, using a mouse to direct a pointer on the screen in order to initiate particular applications or processes, such as printing or saving a file. The user-friendliness of the Macintosh, provided by the graphical user interface and the point and click simplicity of the mouse, together with the availability of packaged software, would obviate the need for programming and typing skills.

The aim of the Macintosh design was therefore to turn the artefact into a tool for individual labour that would be experienced as liberating. It promised the individual 'knowledge worker' freedom from management and central control in the context of the work environment of a large corporation and freedom from dependence on the labour of others in the case of the self-employed 'knowledge worker.'

The metaphor of the Macintosh as an agent of liberation from the drudgery of submission to the standardization and collective uniformity imposed by 'Big Brother' presented in the 1984 advertisement can therefore be seen to have several dimensions designed to appeal to the most deeply felt human aspirations. Perhaps the most important is related to the human need to express creativity through labour, using a tool that is, as much as possible, under the worker's individual control.[5] It promised to free workers in large corporations from the central control that they would experience when using central mainframe systems and the loss of control they would feel when their computer user needs could only be met through the mediation of the data processing department of the corporation where the user-defined problem would be converted into a programme to be run on the mainframe computer. This was necessary because the user would normally lack the technical and programming knowledge required to interact directly with the computer.

The fact that the central mainframe computer in the large company would most likely have been provided by another large corporation, IBM, which then enjoyed a virtual monopoly over the computer market, and was a competitor of Apple, added another dimension to the metaphor. This was the David versus Goliath dimension, with Apple playing the role of David and IBM that of Goliath. The large corporation is thus portrayed as an incarnation of 'Big Brother,' imposing control and uniformity on its workers. IBM was then well-known for its strict code of dress—dark suit and tie—and code of conduct for its employees, in total contrast to the creative freedom existing inside Apple in its early days, containing, as it did, many members of the flower-power hippy generation of California.[6]

The capacity of the computer to facilitate the reorganisation of labour processes through the incorporation into it of many human skills via software, a form of automation, was one of the main reasons for the computer becoming a key technology of the second half of the 20th century. This was first achieved through the mainframe computer, which was costly and only affordable by large corporations. It was also operated centrally and its operation required a high level of technical skills that were not easily accessible to most end users within the corporation for whom the data processing department became the interface between themselves and the machine. Even for workers with technical skills, the control of the machine by the corporation that limited how and when they could use it for their own purposes was often perceived as an unwarranted intrusion into their own autonomy. This identification of the faceless, large corporation with a form of oppression was an integral part of the ethos of the counter-culture that imbued the development of the personal computer revolution. The computer hobbyists who became pioneers of this revolution were strongly motivated to perceive the development of personal computing as a form of liberation from such domination, a feeling that was reproduced in the 1984 advertisement.[7] They met and exchanged their ideas in the Homebrew Computer Club created in 1975 to share computing knowledge. Early members of the club were Steve Wozniak and Steve Jobs, who went on to create Apple. Jobs and Wozniak were also frequent visitors to the Stanford Artificial Intelligence Laboratory (SAIL), another influential centre in the development of the early ethos of the personal computer movement.

The way in which the Macintosh was designed and marketed illustrates several key aspects of the process whereby consumption of technological artefacts is individualised under capitalism. It implicitly highlights key facets of a human being's relationship to both production and consumption and to the nature of power. The artefact is presented as a means to achieve freedom for the individual. The idea of freedom is normally associated with the skill to remove constraints on the individual's ability to fulfil his or her purpose or desire. Usually this freedom is related to the world of leisure.

What is unusual about the way that the Macintosh computer was conceived and presented was that the 'freedom' promised was linked to the world of work and therefore of production, rather than to the more common world of leisure. The Macintosh was being presented to the 'knowledge worker' as 'productivity tool,' enabling him or her to achieve the desired outcome without having to answer to or rely on others, thus gaining a greater control over his or her labour process. As I have already argued, this greater freedom can be understood both in relation to a centralised management control in the context of a large organisation or in relation to the dependence on the skills of others, as in the example mentioned earlier of individual creative workers such as the graphic designers. The attack on IBM implicit in the 1984 advertisement was both an attack on the large corporation that stifled individual creativity and an attack on mainframe computing, the technology that enabled the large corporation to maintain centralised management control over the labour process of its workers. The Macintosh promised freedom for the worker in both respects. It was, therefore, a tool rather than a machine.[8]

However, as I show, the central contradiction at the heart of Apple's message is that in order to deliver what it promised to the consumers of the Macintosh,

Apple itself needed to grow into a huge multinational corporation, controlling the labour of the producers of its artefacts in a way that IBM in 1984 would never have thought possible. This is also a central contradiction of capitalism itself: to promise individual freedom through consumption via the market, whilst in reality its system of production requires workers to be controlled and enslaved.

Production

In the infancy of the microcomputer industry, manufacturing was craft based. The 'inventor's garage' figures prominently in the mythology of the industry, none more so than Steve Jobs's at 2066 Crist Drive, Los Altos, California, where the original Apple I was manufactured. The work, however, involved mainly the assembly and soldering of components bought off-the-shelf from suppliers. The success of Apple II initiated a process of industrial production that gradually over time transformed into a process of mass production. The contradiction between the ideals of contributing to the liberation of labour held by many of the pioneers of microcomputers in Silicon Valley and the reality of employing labour in the highly Taylorised assembly of their products as markets began to grow did not escape Steve Jobs in particular, who tried to maintain as much as possible a humane regime of worker autonomy at Apple. His solution to the conundrum was to seek to automate production as far as possible, hoping, as did the early advocates of automation, that this was a means of freeing labour from the boredom of routinised jobs. The Macintosh was conceived in a separate division of Apple under the close supervision of Steve Jobs. Jobs reputedly paid as much attention to the question of how the Macintosh was to be manufactured as he did to its design and to its friendliness and aesthetic appeal to the user. A new factory for the manufacture of the Macintosh was built by Apple in Fremont, California. When it opened in 1983, the factory was so automated that, according to one report, visitors often outnumbered workers (Smith & Oliver, 1992). An attempt was also made to automate the Apple plant in Carrollton, Texas, where Apple IIs were assembled. The plan involved achieving a daily production of 1,500 computers using only six workers, but it was never put into operation. However, the reality of capitalism soon caught up with Jobs's idealism. His unorthodox management style had already been recognised as a problem, even by Jobs himself, and in 1983 he convinced John Sculley, until then president of PepsiCo., to become Apple's chief executive officer.

Early in 1985, a major recession hit the personal computer industry. In its first attempt to deal with this, Apple closed four of its plants for two weeks (one plant in Fremont, California; a Texas plant where the Apple II was being manufactured; a plant in Cork, Ireland manufacturing Apple IIs for the European market; and a plant in Singapore). In order to minimise the financial effect on employees, they were asked to take their holidays during that period (Rempel, 1985).

However, as the crisis continued, internal tensions developed in the management of Apple as to how to deal with it, particularly between Jobs and Sculley. The situation came to a head in June 1985, when Sculley's views prevailed and Apple underwent a major restructuring that dissolved the separate Macintosh division and closed three of its six plants, with 1,200 employees losing their jobs. A major issue of contention between Jobs and Sculley had been Jobs's opposition to plant closures and job losses. The production of the Apple II was transferred from the closed Texas plant to the highly automated but underutilised Macintosh Fremont plant in California. Manufacture of the latest version of Apple II, the IIe, was transferred entirely to the Singapore plant, which already manufactured it. The restructuring gave Sculley full management control over the company and led to a diminution of Jobs's role. A leading industry commentator at the time remarked that Sculley "was changing Apple from a religion into a business" (Woutat, 1985). Following the reorganisation, Apple abandoned Jobs's focus on automation and sought to implement more of the Japanese management methods of just-in-time and to introduce flexible manufacturing systems (Yoder, 1990; Chisman, 1989).

Jobs was forced to leave Apple in September 1985. He went on to found a new computer company, NeXT, having designed a computer with that name that was never really successful. At NeXT, Jobs helped design a manufacturing plant that was more highly automated than anything that had come before. A *Fortune* report on the plant remarked that the 40-strong manufacturing staff had more PhDs than the group that had designed the computer in the first place (Alpert, 1990. The plant never achieved even 10 percent of its production capacity, as sales of the NeXT computer were sluggish.

Jobs eventually returned to Apple, when Apple took over NeXT in 1996, and became its chief executive officer the following year, a position he used to drive the design and production of highly successful consumer products such as the iMac, the MacBook, the iPod and the iPhone. In the process, Apple maintained a regime of secrecy and tight control over its intellectual property. Apple's manufacturing philosophy changed profoundly in many respects and followed mainstream US brands in outsourcing most of its manufacturing and product assembly operations to subcontractors, mainly in East Asia. Apple has attempted to keep secret the identity of its suppliers and manufacturers. In the run-up to the launch of new products, Apple extracts non-disclosure agreements from its suppliers, which involve such levels of secrecy that any sub-contractor employees who gain knowledge of some aspect of the forthcoming Apple product have to be subjected to intensive surveillance (Pomfret & Soh, 2010).

An investigation by *Mail on Sunday* in 2006 (MailOnline, 2006), of Apple's manufacturing subcontracting, finally began to reveal the huge distance travelled by Apple since its idealistic beginnings in Silicon Valley, ostensibly dedicated to the fight for individual freedom and against monopoly capitalism. The *Mail on*

Sunday investigation discovered that one of Apple's major manufacturing sub-contractors was the Taiwan-based Hon Hai Precision Industry Co. which trades under the name Foxconn Industries, a company that the report alleges employs a million workers and maintains factories in mainland China. Apple product assembly is carried out at two factories. The first factory is in Longhua, near Hong Kong, where the iPod is assembled. The second is in Zhengzhou in central China. A subsequent investigation by *The Wall Street Journal* (Dean, 2007) added further details to the picture. Hon Hai also assembles products for Hewlett Packard, Nintendo, Motorola, Sony, Nokia and Dell. It is the only supplier of Apple's iPhones and a main assembler of iPods. The Longhua factory employs and houses some 270,000 individuals in a fortified city bigger than Newcastle. Workers live in dormitories on the site, 100 to a room. There is no charge for the accommodation, but outside visitors are not allowed. The workers wear uniforms colour coded by department. A *Daily Mirror* reporter described the sessions of 'professional education' that constitute the only break from work routine in the following terms:

> Like soldiers on parade, the young men and women are ordered to line up on the factory roof and drilled for up to three hours, often in searing heat. On occasions they're required to stand still for hours without moving a muscle. These extraordinary exercises were devised to ensure that the workers toe the line. (Webster, 2006)

In an interview with *The Wall Street Journal*, Hon Hai's founder, Terry Gou, said the following: "I always tell employees: The group's benefit is more important than your personal benefit" (Dean, 2007).

All gates are policed by guards and all those entering and leaving are closely checked both to prevent product theft and the leakage of intellectual property. Vehicle occupants are checked with fingerprint recognition scanners. Security guards use metal detectors to search employees, and if any metal is found on an individual, the police are called (Pomfret & Soh, 2010). The *Mail on Sunday* report interviewed one worker who said she worked a 15-hour day for £27 a month (Mail Online, 2006). These are the conditions currently prevailing in the manufacture of the artefacts of the company which in the first few years of its life promised that its products would free workers from the uniformity and oppression imposed by large corporations.

Conclusions

In this chapter, I have tried to highlight the dual condition of human beings as both producers and consumers in the context of analysing the development of personal computing within a capitalist economy. The growth of such an economy requires the transfer of an increasing number of human activities from the realm of subsistence and collective consumption to that of individual consumption via the market. The justification for this is that the individualization of consumption promotes the individual's freedom of choice and frees that individual from externally imposed constraints. Technological artefacts appear as vehicles for enhancing such freedom. Hence "with a Volkswagen you can go anywhere you want to when you want to," whilst a train requires timetable and destination constraints imposed by the need to also consider the needs of others. Normally implicit in these considerations is the assumption that the expression of freedom lies primarily in the realm of leisure.

Work time for human beings is largely perceived as the time during which the means to meet basic needs and to enjoy the freedom provided by leisure are obtained. This perception was already present at the birth of the age of mass consumption, when Henry Ford introduced the US $5 day for his car workers partly as a compensation for the fact that his factories turned them into repeating machines but also so that they could enjoy the products of their labours, the model T Fords, in their leisure time: "High wages to create large markets," Ford said (Grandin, 2009). And he promised that his products would provide freedom in leisure.

The history of personal computing is a case study in how an innovation inspired by the desire of human beings to free their creative labour from the stifling control of monopoly capitalism gradually became a new instrument of corporate control and oppression of labour unable to escape the dynamics of the system itself, now dominated by new monopolies that they themselves created. The pioneers in the early days of Silicon Valley, the members of the Homebrew Computer Club and of SAIL, including Steve Jobs himself, believed in a code, according to which

- Access to computers—and anything that might teach you something about the way the world works—should be unlimited and total.
- All information should be free.
- Authority should be mistrusted and decentralization promoted.
- Hackers should be judged by their hacking, not by such bogus criteria as degrees, race or position.
- Art and beauty can be created on a computer.
- Computers can change your life for the better.[9]

The members of the Homebrew Computer Club wanted everyone to have access to the wonderful tool they thought the computer to be. The organisation that gave birth to the Homebrew Computer Club was called the People's Computer Company. But their original vision was one of giving computing skills to everyone in order to enable them to reconfigure the machine to their own purpose, rather than removing the need for technical skill from the use of computers. They wanted the production of personal computing to be part and parcel of its use and did not conceive technical skill to be a barrier to the democratisation of computing. The spirit of the Homebrew Computer Club was one of cooperation and of sharing of possessions and knowledge, not one of competition, secrecy and surveillance. They opposed the increasing division of labour that would lead to the de-skilling of the labour process of production and a concept of freedom that was relegated entirely to the realm of leisure.[10] That same spirit has subsequently resurfaced in the world of computing, through such movements as the early stages of the development of the World Wide Web and, later, the open source software movement.

However, the logic of capitalism dictates otherwise. The growth of the market necessitates the increasing separation of consumption from production. In order to achieve this, technological artefacts have to become 'user-friendly.' In this process, the user loses all understanding of how the technology works and the technology becomes a 'black box.' The realm of technology becomes a mysterious one for the user-citizen, who feels alienated and unable to exercise any sort of control.

The growth of the market, as I have shown, also requires the development of mass production. This leads to an increasing division of labour in production, the de-skilling of workers and/or their replacement by machinery to increase productivity. However, mass production also reduces price. This democratises consumption insofar as the artefact becomes accessible to people of more modest means and with lower levels of technical knowledge. But mass production, as I have shown in the case of personal computers, also inexorably leads to the oligopolisation and monopolization of production.

The Macintosh democratised the consumption of computing through its radical reduction in the level of technical skill required for its use. Its mass production also enabled it to be economically accessible to a larger number of people. In this sense, the metaphor of the 1984 Superbowl advertisement might seem justified. However, the cost is the increasing subjugation and alienation of the individual in the realm of its production. Steve Jobs, as an early adherent of the hacker ethic and member of the flower-power generation, was aware of this. His solution for the conundrum was the almost complete replacement of human beings by machines in production, full automation, an objective he pursued relentlessly in his early days at Apple and subsequently in NeXT. Freedom of the individual would be guaranteed by the removal of the necessity for his or her participation in production. In this respect, his vision was shared by Karl Marx (1894/1981):

> The realm of freedom really begins only when labour determined by necessity and external expediency ends; it lies by its very nature beyond the sphere of material production proper . . . . Freedom in this sphere (of production) can

consist only in this, that socialised man, the associated producers, govern the human metabolism with nature in a rational way, bringing it under their collective control instead of being dominated by it as a blind power; accomplishing it with the least expenditure of energy and in conditions most worthy and appropriate for their human nature. But this remains a realm of necessity. The true realm of freedom, the development of human powers as an end in itself, begins beyond it, though it can only flourish with this realm of necessity as its basis. The reduction of the working day is the basic prerequisite. (pp. 958–959)

However, the logic of capitalism did not permit Jobs to realise his vision. In order to prevent markets from becoming saturated and growth to stop, artefacts have to become obsolescent in as short a time period as possible, a process that is facilitated by continuous innovation and leads to the creation of landfill mountains of PCs. In the case of the information technology industries, this is guaranteed largely by the rate of technological change in the semiconductor industry, by the pursuit of Moore's 'law.'[11] When market saturation threatens, the pressure intensifies to customize the artefact ever more finely to perceived individual consumer requirements in such a way that the consumer will feel the need to jettison the artefact he or she already possesses. Both of these processes militate against full automation, as they require flexible production processes and shorter production runs, which limit the time that companies need to recoup their investment in machinery. In any case, machinery is not as flexible as people, who are also the sole source of profit in the full value chain. The requirement to reduce price in order to extend the market and facilitate economies of scale also leads to a pressure to reduce wages. The extreme suppression of the freedom of Chinese workers who are paid a starvation wage assembling Apple's products in China by Foxconn Industries is, at least partly, a consequence of the market needs of Apple. These include the ruthless protection of the company's intellectual property rights, in direct contradiction to the hacker ethic so beloved of the young Jobs.

Since the 1970s, in countries that are part of the OECD, the value added by production has tilted in favour of capital and against labour (Lübker, 2007). In order to continue to maintain their consumption levels, workers have had to work longer hours. Between 1970 and 2008, the average number of hours worked per year by a worker in the United States increased by 20 percent.[12] In those countries where workers succeeded in reducing the length of their working week, such as France, the gain has been under threat, particularly since the 2008 credit crunch and through the subsequent measures taken by the government to deal with its effects. The pressures to increase the age of retirement, which have grown in recent years, create a further threat to leisure as the road to freedom.

Thus, in the recent past, the ability of individuals to achieve freedom in their leisure time through consumption via the market has become increasingly limited. Their freedom at work, both in the West[13] and in the sweat shops of China, has also been increasingly curtailed.

However, the spirit of the pioneers of Silicon Valley survives and keeps resurfacing in different guises, despite all the attempts to subject the whole of human life to the dictates of the market and of the accumulation of capital. It is the spirit that Marx (1894/1981) expressed when he stated that freedom in production could only be achieved when

> socialised man, the associated producers, govern the human metabolism with nature in a rational way, bringing it under their collective control instead of being dominated by it as a blind power; accomplishing it with the least expenditure of energy and in conditions most worthy and appropriate for their human nature. (pp. 958–959)

Whether his faith that capitalism would develop technology that would enable 'real freedom' to be achieved in the realm of leisure through the reduction in the length of the working day once the 'associated producers' brought it under their collective control is justified seems, at the very least, questionable.

Endnotes

1. Karl Marx described this as the 'object of labour,' to be distinguished from the 'instrument of labour,' that is, the tool. The 'object' of labour is the aspect of nature that the human being works on to achieve his or her purpose, whilst the 'instrument' of labour (or tool) is the implement used to achieve that purpose.

2. Marx defined a tool or 'instrument of labour' as "a thing, or a complex of things, which the worker interposes between himself and the object of his labour and which serves as a conductor, directing his activity to that object" (Marx, 1857/1976, p. 285). He further pointed out that a good tool should not require the user to be aware of the existence of the toolmaker: "It is by their imperfections that the means of production in any process bring to our attention their character of being the product of past labour. A knife which fails to cut, a piece of thread which keeps on snapping, forcibly reminds us of the Mr. A, the cutler or Mr. B, the spinner. In a successful product the role played by past labour in mediating its useful properties has been extinguished" (Marx, 1857/1976, p. 289).

3. The term *PC* is generally used in an ambiguous fashion. Sometimes it is used as a generic term for all personal computers; other times it is meant to refer to a personal computer that complies with the standard originally set by the IBM PC. Here the term is employed exclusively in the second sense. The full expression, 'personal computer,' is used when I wish to refer to any computer designed for individual use, whatever its architecture. At the time of writing, this would cover almost exclusively PCs and Macintosh computers, as other architectures are no longer available in the market.

4. The concept of the graphical user interface was first developed by Xerox at its Palo Alto Research Centre. Steve Jobs visited this centre in 1979 and was inspired by the visit to develop a similar interface for the Macintosh.

5. Marx referred to this fundamental characteristic of human nature in the following terms: "Labour is first of all a process between man and nature, a process by which man, through his actions, mediates, regulates and controls the metabolism between himself and nature. He confronts the materials of nature as a force of nature. He sets in motion the natural forces which belong to his own body, his arms, his legs, head and hands, in order to appropriate the

materials of nature in a form adapted to his own needs. Through this movement he acts upon external nature, and changes it, and in this way he simultaneously changes his own nature. He develops the potentialities slumbering within nature, and subjects the play of its forces to his own sovereign power . . . . Man not only effects a change of form in the materials of nature; he also realizes his own purpose in those materials" (Marx 1867/1976, pp. 283–284).

6. Steve Wozniak, cofounder of Apple with Steve Jobs, said of Jobs: "Steve was into everything hippy, he ran around shouting 'free love man' and eating seeds as he embraced the flower power set" (Neate, 2008).

7. Ridley Scott was chosen to direct the '1984' Macintosh advertisement probably because he had, two years previously, directed *Blade Runner*, the film that Theodore Roszak described as having "most effectively exposed the decadent underbelly of corporate high tech" (Roszak, 2000).

8. Marx differentiates machinery from tools. Whilst tools are simple aids to labour, typical of craft production, a machine is "a mechanism . . . that after being set in motion, performs with his tools the same operations as the worker did with similar tools" (Marx, 1867/1976, p. 495). The idea of automation of labour is, therefore, an integral part of Marx's concept of machinery, and he views machinery, defined in this way, as typical of industrial production. The main-frame computer can be perceived in Marx's terms as a machine able both to automate and to control work. Marx did not envisage, however, the possibility that a tool could itself automate the labour of another worker. This is, as I have shown, what happens when the Macintosh computer, with a desktop publishing package, is used by a graphic designer.

9. This was the 'hacker ethic' as described by Steve Levy in 1984, cited in Markoff (2005, p. 96).

10. See Hauben (1995) for a discussion of the radical nature of the early personal computing movement.

11. In 1965, when the semiconductor industry was still in its infancy, Gordon Moore, a cofounder of the industry's leading firm Intel, predicted that the number of transistors that could be placed on a chip would increase exponentially, doubling every year (Moore, 1965, p. 4). In 1975, he revised this prediction to every two years (Moore, 1975).

12. Figure quoted by Wolff (2009).

13. In this respect, see, for instance, Head (2003).

THREE

The ICT Value Chain

Perpetuating Inequalities

Richard Sharpe

ICTs are liberating: they allow their human users to access seemingly ever widening ranges of information, of services and other people. They are increasingly mobile, liberating the user from the end of a fixed-line for communications. They, to use the term of Frances Cairncross (1997), kill distance. The physical nature of humans means that we have to be in one place at a time, not necessarily close to the information, services, or people with whom we wish to make contact. Mobile ICTs provide easier access to emergency services; they inform as never before. All this is not confined to North America and Europe. For example, China is a growing market for ICT devices. There is a stream of uplifting stories in the media about users of mobile ICT devices in rural areas who benefit from their increased knowledge of markets, prices and conditions as a result of their use of ICT devices. The use of ICTs can often be beneficial. But this chapter examines another vital issue, which is not considered enough: how does the extraction of vital elements for the production of ICTs and other complex commodities contribute to the inequalities between humans? A rapidly growing set of ICT devices and services with an increasing range of applications is used by increasing numbers of people in growing numbers of locations in the world. Yet, given the benefits from the uses of ICTs, what has to be put on the other side of the equation in terms of the extraction of vital materials such as gold, which I focus on in this chapter? Who pays for the benefits? Who are those who pay and where are they?

Compare the position of the child in the many countries where artisanal and small-scale mining of gold is an important part of that child's life, because it generates income for the family, with its associated health risks, with the young person in another part of the world, where the use of ICTs is essential to that child's life: In the former, the child is a provider of an essential part of the device used and may well be damaged by the processes of extraction; in the latter, the device and service are an essential part of the child's social life, not a want but so essential that many feel addicted. If we just look at the take-up of smartphones to 2011 in the UK, we can see how many ICTs are woven into people's lives. Just under half of all teenagers had a smartphone by the time of Ofcom's 2011 (Ofcom, 2011a) report on markets. Of those teenagers with a smart phone, 60 percent of them felt addicted to smartphones. "Our research into the use of smartphones, in particular, reveals how quickly people become reliant on new technology, to the point of feeling addicted" (Ofcom, 2011c).

In this chapter, I examine one of the first steps in the value chain, the extraction from nature of one vital material, gold, and conclude that labour and economic processes help sustain and reproduce long-established inequalities.

Value Chains

The production of ICT devices, as of other complex commodities—from semiconductors to mobile phones to motor cars—is one step in long value chains that start with extraction from nature and end with disposal. These have been called global value chains to emphasise the global nature of production (see, for example, Humphrey & Schmitz, 2001). They have also been called *systems of provision* (SoPs): SoPs "are the inclusive chain of activity that attaches consumption to the production that makes it possible" (Fine, 2002, p. 79). I prefer the term *value chains,* as this reflects the buildup by human labour of value in a commodity. The importance of a global value chain for a particular sector is shown in world trade figures for manufactured intermediate goods (MIGs), an incomplete product that is shipped to another location for further production. Electronics, of which ICTs form a large part, is the largest product group in world trade in which MIGs form a part. A total of 43 percent of the top 50 industries and products with global trade in MIGs was attributed to electronics in 2006 (Sturgeon & Kawakami, 2010, p. 248).

Most of the work on global value chains has, as the work of the Global Value Chains Initiative, focused on the material products: where they were manufactured, where they were passed for the next stages of manufacture, and what were the manufacturing steps in each location. This 'classic' view of the global value chain needs to be extended to the outer edges of production and disposal of a product and to include not only how the product is made in the different links of the chain but also who is making it. By this expansion, we will see where and how inequalities among people are generated, by including humans in their productive roles and in their locations.

The products of this value chain enter, after successive steps, into a market, a final market where commodity is complete or near complete and can be purchased by a buyer for its use, rather than for turning it into yet another commodity. The consumer in these markets focuses on the balance between exchange value and use value: will consumers pay this or that for this or that commodity? The physical commodity they purchase, however it is purchased, with its software embedded, does not demonstrate to the consumer the different labour processes by which the extracted natural materials in this product in its packaging in their hands was recovered from nature or how they were transformed in the successive links in the chain. Neither does it easily show where these different labourers were located as they performed their labour to transform what were original materials gained from nature into the product that faces them. The consumer does not know, often cannot know, where, for example, the gold came from, what the conditions of work were or the degrees of political, economic, gender, or generational oppression or freedom under which those individuals worked. Physical objects normally bear the 'country of origin' mark. This is now obligatory in North America and the Europe. However, this tells the consumer nothing of the origins of the raw materials or where the components—the manufactured intermediate goods of which the product is made—within the device were originally made. It only tells the consumer of the last location of assembly. Consumers, focusing on the balance of what they will pay and how they can use the commodity, are shielded from this information, unaware that they are but a step in a long and complex value chain that links human to human through the curtain of capitalist commodity markets starting at extraction, humans who are unequal in this value chain. Within each step of this long value chain there are inequalities: inequalities of location, reward for labour, and the impact the labour process has on the worker. The salary for a mining engineer in South Africa in late 2012 was R120,000 ([£8,608]; Payscale 2012). At the other end of the value chain, a survey showed that the average salary for permanent IT professionals in the UK was £38,946 in 2011 (Williams, 2011). Moreover, the working conditions of the gold miner are appalling compared with those of the IT professional. These are just two points in the global division of labour that capitalism has generated. Sussman and Lent (1998) stated the following:

> Pursuing the logic of a singular, integrated, and telecommunications-linked world economic system, [Trans National Corporations] and their state allies have mobilized workers into a low-wage, segmented, and flexible global production force, made up of men and women who for the most part will never have the purchasing power to enjoy the goods and services created with their own labour. (p. 2)

This long global value chain is true of all physical commodities. Given limitations of space, I shall only be able to focus on the first steps in this value chain in relation to ICT devices—the extraction of materials from nature—and only on one material extracted—gold.

Extraction

An essential part of the production of ICTs is the small but vital amounts of precious metals needed in their construction. Take the physical components of the mobile phone, now a high volume ICT device. Nokia, in an attempt to be ethical in its production, and to appeal to ethical consumers, breaks down the physical components of its phone ranges:

Plastics	45%	(thermo plastics and thermo sets)
Metals	35%	(copper, iron and aluminium alloys)
Glass and ceramics	10%	
Battery electrodes	9%	(graphite, lithium and cobalt)
Precious metals	0.11%	(gold, silver, palladium and platinum)
Other	0.9%	

(Source: Nokia, 2010)

Nokia and other makers of ICT devices would not use gold unless absolutely necessary. Gold is used because it is a good conductor that will not rust or otherwise corrode. It is used not only in mobile telephones but also in all of the network computers, routers and other devices where gold's properties as a noncorrosive and very effective conductor are needed. Within every server, PC, mobile device and so forth, there is gold, likewise within every router, base station and in the very guts of the networks on which the mobile ICT device depends. It is within every device that accesses these networks, within every laptop and netbook, within every iPad and so on. About 3,300 tonnes of gold are mined a year. "In all of history, only 161,000 tons of gold have been mined, barely enough to fill two Olympic-sized swimming pools. More than half of that has been extracted in the past 50 years" (Larmer, 2009, p. 43). At current values, it is worth about £72 billion a year. The largest use of gold is jewellery at about 50 percent. The second largest use is as a store of value: mainly gold bullion (World Gold Council, 2011, p. 28). Much of this gold is mined by large-scale mining operations. Electrical devices, including ICT devices, are the third largest users of gold. A total of 11.7 percent of all gold by market value mined in 2010 was used in high technology devices.

The current structure of the world gold mining industry was formed in the 1980s and 1990s when countries with gold deposits were urged by those who advocated the neoliberal globalisation agenda to lower restrictions on mining and, if they had any, privatise their nationalised mining operations. These gold-holding countries were promised increased foreign investment, higher production and rising employment as a result: They would also be earning more foreign exchange (Twerefou, 2009, pp. 15–16). The beneficial impact of this reformation of the gold industry has been questionable. Large multinational mining operations have negotiated the rights to mine extensive tracks of the gold-holding countries. They are allowed to keep at least 75 percent of their financial resources outside the countries in which they mine in order to pay for capital equipment, debt servicing, payments to expatriate employees and dividends. This has enabled many mining companies to retain most of their earnings in offshore accounts, leaving very little in local accounts to cover local operating costs. The little left for local purchases in real terms could be very minimal in some countries, since the import content of local purchases such as petroleum products, explosives and other goods are not taken into account. Thus, while mining contributes substantially to gross foreign exchange, the real benefits accruing to the nation may be very little (Twerefou, 2009, p. 17).

The world's five largest gold mining corporations are as follows, in order of market capitalisation at the end of 2010: Barrick (Canada), Gold Corp (Canada), Newmont (USA), Kinross (Canada), AngloGold Ashanti (South Africa), and Gold Fields (South Africa/UK) (Barrick Gold Corporation, 2010, p. 46). The world's two largest gold mining companies by market capitalisation, Barrick and Gold Corp, pay more in dividends to their shareholders than in royalties to the countries whose gold they mine. Gold Corp paid $154.4 million in dividends in 2010 against $109.7 million in royalties. For the three financial years from 2008 to 2010, Gold Corp paid a total of $414.8 million in dividends to its shareholders and $246.3 million in royalties (Barrick Gold Corp, 2010, pp. 35, 72). Barrick paid $436 million in dividends in 2010 and $287 million in royalties (Barrick Gold Corporation, 2010, pp. 69, 123). Where do these dividends go? The 2010 annual report of AngloGold Ashanti (AGA) shows us. AGA is based in South Africa where it is registered. The bulk of its operations are also in South Africa, with six deep mines and one surface mine employing 57 percent of AGA's total workforce worldwide and extracting 39 percent of the gold AGA extracted worldwide in 2010. But the majority of the shareholders are in the United States. Just over one fifth accrue to shareholders in South Africa. All of these shareholders took $131 million out of the company in 2010 dividends. One in ten is in the UK where AGA has no operations at all. To pay these dividends to capital, AGA had to raid its reserves, as its profit in the year was $129 million.

As we can see from Table 2, of the costs of production to Newmont (the third largest gold mining corporation by market capitalisation in 2010) for an ounce of gold in the three financial years, 2008 to 2010, royalties and production taxes charged by the nations in which Newmont mines gold are a small proportion of the overall costs of production. They are always in these years smaller than the price Newmont gains for the other minerals it is able to extract from the gold ore,

38 | Richard Sharpe

which it calls by-product credits. They are often copper or silver (Barrick Gold Corporation, 2010, p. 103).

Table 3.1: The Location of Shareholders of AngloGold Ashanti

Country	Percentage of Shareholders
United States	52.60%
South Africa	22.54%
United Kingdom	11.73%
Ghana	2.95%
France	2.35%
Rest of Europe	2.56%
Rest of Americas	1.20%
Rest of the world	4.07%

(Source: AngloGold Ashanti, *2010 Annual Report*, p. 9)

Table 3.2: Production Costs of a Gold Ounce

Production costs per ounce sold:	2010	2009	2008
Direct mining and production costs	$493	$418	$432
By-product credits	(39)	(30)	(25)
Royalties and production taxes	26	20	18
Other	5	3	4
Costs applicable to sales	485	411	429
Amortization	26	105	103
Reclamation and remediation	6	4	4
Total production costs	$617	$520	$536

(Source: Newmont Mining Corporation, 2010, p. 64)

Table 3.3: Royalties as a Percentage of Production Costs and Sales Price in 2008–2010

Year: Royalties	Production Costs	Sale Price	Royalties/ Production Costs	Royalties/Sale Price
2010: 26	617	1222	4.21%	2.13%
2009: 20	520	977	3.85%	2.05%
2008: 18	536	874	3.36%	2.06%

(Source: Newmont Mining Corporation, 2010, pp. 43, 64)

Table 3 reveals the relationship between royalties and mining taxes and production costs and the price Newmont got in the market for an ounce of gold averaged for that financial year. We see in these three years that royalties are a small fraction of total production costs and barely get over 2 percent of the price per ounce realised by Newmont in a year. There is, therefore, precious little that the gold-owning nations get for their gold, compared with the shareholders of the gold mining corporations.

I have shown what these mining companies pay to capital (dividends) and to the countries whose gold they extract (royalties and mining taxes). These companies are more secretive about what they pay labour. One of the gold mining companies that claims to be in the top 10 worldwide, Harmony, has revealed its employee salaries (including directors and management), retirement and other benefits (excluding employee tax). This was R4,193 million in 2010 (Harmony Gold Mining Company Limited, 2010, p. 354). That translates to $14,062 per employee. This includes the salaries of managers and directors as well as their pension contributions so that the actual salary for a gold miner will be considerably below this annual figure. Harmony mostly operates in South Africa but it has opened mines in Papua New Guinea.

The South African National Union of Miners said in 2011 that gold miners in South Africa earned, in the first half of 2011, on average, 3,800 Rand (£346) a month: £4,152 a year (BBC, 2011C). This was before a series of strikes in the summer of 2011. They won an 8 percent increase to an average of £4,484. Inflation in South Africa was then running at 5.3 percent, meaning that their real increase was under 3 percent in actual spending terms. Little local employment was created by the introduction of mining corporations into gold-holding countries because, first, the major mining companies use capital-intensive methods of extraction and not labour-intensive ones, and second, the labour they do hire to operate their machinery is often from outside the country as those within it do not have the skills to operate the machinery. A report for the United Nations stated the following:

> The contribution of mining to employment generation is mixed and mostly marginal. While real incomes in the mining sector appear to be higher than the national average in many countries, the overall employment impact is limited compared to other sectors such as industry, services and agriculture due mainly to the capital-intensive nature of mining operations. Mining also displaces a large population and denies them access to land, an important employment security for rural people. Indeed, some have argued that the net employment effect of mining is negative given the massive displacement of small miners to marginal sites as well as the abandonment of agriculture as a source of livelihood by many rural communities. (Twerefou, 2009, pp. 19–20)

As the gold on the surface of the earth has been mined, so men, women and children have to go deeper into the earth to extract it. Much of this mining is in countries with loose health and safety regulations, such as Chile or Ghana.

Some work under the most hazardous conditions, as was revealed on August 5, 2010, when 33 gold and silver miners were trapped 700 meters below ground for months in the San Jose mine at Copiapo in Chile (BBC, 2010b). Less publicity was given to the death of six miners in a Venezuelan gold mine in the same month (Reuters, 2010). In Australia, another significant source, there were gold mining disasters leading to deaths in 2000 and 2006. These are only a few of the recorded incidents.

A quarter of this gold is mined by poor migrant workers in artisanal or small-scale mines who make up 90 percent of the workforce. According to the United Nations Industrial Development Organisation (UNIDO), there are between 10 million and 15 million so-called artisanal gold miners around the world, from Mongolia to Brazil. Employing crude methods which have not changed in centuries, they produce about 25 percent of the world's gold and support a total of 100 million people (Larmer, 2009, p. 44).

The conditions of these miners are even more fraught than the gold miners employed by large multinationals. In Ghana in 2009, for example, 18 people, including 16 women, died when a small-scale, unregulated mine collapsed. In June 2010, 32 were feared dead again in Ghana when an unregulated mine collapsed (Adadevoh, 2009). The majority of these accidents go unrecorded because of the itinerant nature of the workforce and the often 'illegal' areas in which they work, that is, areas allocated to the mining corporations by the host country in which the indigenous people have no legal right to mine. The miners get, on average, £156 a troy ounce (£5 a gram) for the gold they mine. They sell it to dealers who get about £8 a gram. In comparison, the price on the world market for gold was as much as £867 a troy ounce at that time (McDougall, 2010, p. 26). The miners get 18 percent of the price of the gold despite the dangers of their labour and the possible damage to their health; yet the dealers, who risk nothing apart from the shift in prices, get 60 percent more.

As Map 1 shows, in almost all of Africa and parts of South America, from 2 percent to over 20 percent of the total population in those regions are dependent on artisanal and small-scale mining, much of it for gold, according to the Communities and Small-Scale Mining initiative hosted at the World Bank (Communities and Small Scale Mining, 2012). Battles over the control of these mines often lead to violence, particularly in the Democratic Republic of the Congo. Artisanal miners often mine on the lands that have been granted by host nations to the large mining corporations. They then come under attack from the security forces of the companies and the state police (Simpson, 2010).

These artisanal miners use mercury to extract the gold from the ore. This leads to systematic poisoning in gas and liquid forms. Mercury enters the water supply, affecting not only the miners but their estimated 100 million dependents. "Environmentalists estimate that, for every gram of gold extracted, miners dump two or three more times mercury into the air and water" (Colline & Phillips, 2011, p. 25). "UNIDO estimates that one third of all mercury released by humans into the environment comes from artisanal gold mining" (Larmer, 2009, p. 44).

Map 3.1 Population Economically Dependent on Artisanal and Small-Scale Mining Around the World

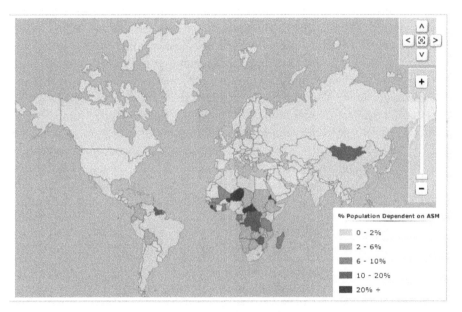

Source: Communities and Small Scale Mining, 2011
Reproduced with permission of Artisianalmining.org

Mercury is a very toxic metal. Its vapour lasts about a year in the atmosphere. Inhalation of large amounts can damage a developing foetus by creating severe developmental disabilities in the child through widespread damage to the foetal brain. It can cause delays in walking and make the child photophobic and have joint pains. In children and adults, mercury will target the kidneys. It can cause acute corrosive bronchitis, lung disease and central nervous system affects including tremor or increased excitability, memory loss and insomnia. It can, in sufficient quantities, be fatal, even to adults (Klaassen, 2008, pp. 947–950).

The consumer of the ICT device and service does not see this labour, either in large-scale mining or in artisanal mining of gold, or the conditions under which it works, or the impact it directly has on these workers' health or on the general environment. The consumer is separated by the market for the device from this realisation of the conditions of those who have laboured, in this case, in extraction to bring them the device. The market only transmits information on price, functionality and brand. Marx (1998, p. 131) stated the following: "Every commodity is a symbol, since, in so far as it is value, it is only the material envelope of the human labour spent upon it." But the social and economic circumstances of this labour embedded in the commodity and by which it creates value are inside the 'envelope' of the commodity and not observable by the commodity buyer. The market does not carry information about labour.

Some ICT device makers try to source their precious metals ethically by buying from certified suppliers. Nokia, for example, tries to implement a policy of buying ethically (Nokia, 2011). But there can be between four and eight levels in the supply chain of materials before it reaches Nokia: Nokia cannot really know the origin of the gold Nokia has to use. Manufacturers of ICT devices depend on the adherence of all levels in this supply chain to substantiate claims that they buy ethically. But this supply chain can never guarantee that the metals being supplied are ethically extracted. Once a metal is refined, its origins depend on paperwork that can be falsified. Even a simple low-value commodity like eggs attracts fraud (BBC, 2010a). The profits to be made from precious metals are far higher than for eggs, thus attracting more imaginative fraud. The Fairtrade Operation and the Alliance for Responsible Mining launched Fairmined in 2011. This is to certify the origin of gold so that it is sourced ethically. The certification scheme has five aims:

1. To get a better price for the miners
2. To give the miners a social premium of 10 percent of the London fixed price
3. To encourage miners to collaborate and form groups
4. To mine safely and responsibility when it comes to the impact on the environment
5. To help bring peace to areas where conflict is generated by gold mining. (Fairtrade Foundation, 2011)

At the time of writing, none of the users of gold in ICT devices has signed up in agreement with the Fairmined principles. Fairmined is initially aimed at the jewellery market and aims to have 5 percent of that market in fifteen years (Fairtrade Foundation, 2011). The whole Fairtrade model, of which Fairmined is a part, has its critics, one of the most trenchant being Sidwell (2008). It is unfair, he argues, because it offers only a very small number of farmers (read miners) a higher, fixed price; it operates to keep the poor in their place, sustaining uncompetitive farmers (read miners), holding back diversification, mechanisation and moves up the value chain; and just 10 percent of the premium consumers pay goes to the producer because retailers pocket the rest (Sidwell, 2008, p. 3). Other critics cite the fact that in cocoa production in West Africa, the Fairtrade certified buyer accounts for a decreasing proportion of purchases in the Ghanaian cocoa market. The reasons are that other buyers also offer incentives just as valuable as the Fairtrade offering: equipment, pesticides, cash payment and credit for the inputs farmers need. "In this rush for cocoa, Fairtrade means little" (Ryan, 2011, pp. 98–99), despite the celebrity backing and visits by Western politicians endorsing Fairtrade.

Fairmining is not the only initiative aimed at creating equality in markets and labour processes. There are numerous other national, regional and global bodies that have taken up the issue of the artisanal mining community of the world. These include the Alliance for Responsible Mining based in Colombia, which is supported by NGOs and pro bono work by the private sector. There is also the Communities and Small-Scale Mining initiative, which was formed in 2001 by the World Bank and is currently chaired by the UK Government's Department for International Development, with some funding help from France, Canada and other nations. Consumer groups also point to the issues involved: for example, Supergreenme, a privately-funded initiative by environmental enthusiasts, points to the impact artisanal gold mining has on those involved (Supergreenme, 2009). Despite all of these efforts and many more, the inequalities in gold extraction are continually reproduced every day, given the power of the forces that continually recreate inequalities.

Even if the device makers could guarantee the supply of their precious metals, they still benefit directly and economically from the artisanal labour used to extract a proportion of these metals: The price is correspondingly lower the more artisanal labour is used to extract it. Withdraw the quarter of gold from the market that artisanal labour extracts and the price goes up significantly for the ethically-sourced precious metals. This artisanal labour, unlike the gold it extracts and the profits the merchants of gold make on it, cannot legally cross borders. The product of this labour and the profits from it can, as a result of globalization, move across the globe into almost any country that those who own it choose. The labourers and their dependents do not have this freedom. This is an inequality recreated between people as a result not of their work but of their birth. The globalized economy allows rich people to move across the globe, while those who only have their labour cannot do so easily, especially if that labour is artisanal in the extraction processes.

Conclusion

If we are to analyse the inequalities sustained by the manufacture, consumption and disposal of ICT devices and other complex commodities, then we must take a much broader view than the traditional one of focusing exclusively on production and consumption. I have focused on the extraction from nature of the essential material of gold. This is because, of all the raw materials taken from nature, gold has the most extensive literature.

Consumers buy their commodity balancing out use and exchange—utility versus price. The commodity 'appears' on the market promoted by the vendor with a focus on its utility. The market provides a curtain partly shielding the consumer from knowledge of the extraction and production phases in which inequalities are continuously reinforced. The inequalities I have identified as being sustained in the value chain of the production of complex commodities are twofold. First, there are the inequalities between capital ownership and reward and the people of the nations where the gold is to be found as clearly shown in the difference between royalties and dividends. Second, there are the inequalities of labour in gold extraction compared with others involved in the production of commodities as a result of the privilege given to certain labour processes and of the funding of these extraction processes. There are 100 million people involved in artisanal production whose income is low as a result of the take by merchants and whose health is in jeopardy as a result of the extraction methods they have to use. Even for those in organised labour, the pay granted by dominant global mining companies is low.

The mechanisms by which these inequalities are sustained and reproduced day by day are twofold. First, there is a continued growth in the consumption of ICT devices that is required in capitalism to keep profits rolling. Second, the difficulty of inserting ethical considerations into the purchase of commodities and services, because of the role of the market as a curtain, blinds the consumer to the generation of inequalities in this value chain. Manufacturers can play a pivotal role as many try to 'ethically source' their raw materials and subsystems and increasing numbers of individual consumers attempt to buy ethically. But the persistent sustenance of inequalities in the whole value chain continues, despite these efforts. Such inequalities can be changed by contest. How much should unionised gold miners get for their labour? How much should share owners of a gold mining company get in dividends? How much should artisanal miners get for their product, compared with the merchant? Where should the artisanal miner be able to mine? How much should the nation that has gold deposits on its land get, compared with the capital it has called into its country to extract that gold to be used in the production of today's ICT devices?

The low pay of miners and the appalling conditions in which they work are continuously created primarily by corporations maximising their profits to meet the demand for gold and by the need to sustain the 100 million supported by artisanal mining.

Alvaro de Miranda (2009) has suggested that the concept of the *information society* has been consistently used to create "dreams of a future techno-utopia in which all will participate, from which all will benefit, where the deepest human aspirations will be fulfilled"(p. 34). But the creation and diffusion of the information society depend fundamentally on the ever increasing production of ICT products. Consideration of one stage in the supply chain of ICT production has demonstrated conclusively that, so far, the diffusion of the information society has been dependent on the perpetuation of appalling inequalities in terms of the pay and conditions of gold miners. Numerous organisations, NGOs, private and government initiatives and intergovernmental organisations such as the OECD regularly or occasionally report and act on the position of artisanal and small-scale gold mining operations in the world and how they recreate inequalities across the globe. However, they are unable to demonstrate that their efforts can change the general pattern of the reproduction of inequalities.

FOUR

Climate Change, Industrial Animal Agriculture and Complex Inequalities

Developments in the Politics of Food Insecurity

Erika Cudworth

This chapter examines changes in agricultural practices and technologies, focusing on developments in meat production. There are a number of interlinked processes that are examined, beginning with the historical development of livestock farming, and the impact of these processes and of very recent developments in animal food production on local, regional and global environments. The current scale of animal farming is both extensive and intense, and it has been growing rapidly since the 1950s. As a result, there has been a dramatic increase in the populations of farmed animals. In 2003, for example, the United States became the first country to raise over one billion farmed animals in a single year; this was more than twice the number of animals raised for food in 1980 and 10 times the number raised in 1940 (Marcus, 2005, p. 5). Since 1980, global meat production has more than doubled, but in the South (where levels of meat and dairy consumption are rising every year), it has tripled. Sixty billion animals are currently used each year to provide meat and dairy products. Considering current trends, this figure could reach 120 billion by 2050 (MacDonald, 2010, p. 34). The UN Food and Agriculture Organization predicts a dramatic rise in human population to 8.9 billion by 2050, and the rise in the food animal population is promoted partly by this increase and also by heightened demand in both richer and poorer regions of the globe (Giles, 2009).

The production of animals and animal feed crops has had a significant impact on localised food production systems, and the intensive production of stock is set to become the model for agricultural development in poor countries. Animal-based food is seen as a solution to food poverty and helping to eliminate food insecurity. This chapter suggests however, that the establishment of Western intensive production and the promotion of Western eating habits are more likely to increase social inequalities and insecurities than to increase food security. In addition, these increases in meat production and consumption are highly likely to have 'disastrous' impacts on the environment, and in particular, in terms of an increased contribution to climate change (Goldenberg, 2012).

Global Markets and Industrial Animal Protein

This section traces the ways in which the current global system of animal-based food production was tied in to developments in technology and bound up with preexisting and newly emerging forms of social and economic inequality. It maps the changes in the farming of livestock and the production of animal protein in Europe and North America. It traces the shift from a localised system of production, through to the specialisation, intensification, integration and mechanization of the livestock industry. The operations of local, regional and global networks of relations shaped the development of animal food production, and the production and consumption of meat are historical processes in which systemic relations of species are constituted with and through relations of capitalist colonialism.

From Local to Global: Class, Colonialism and Species

In many pre-industrialised countries, domesticated animals have often been more important as a means of labour power than as a source of food. The work of anthropologist Marvin Harris (1987), for example, has indicated that throughout India, cattle were the principal means of ploughing, provided an important form of transport, and were deployed in various agricultural processes, such as winnowing and flour making. Even pigs, whose flesh currently constitutes 40 percent of global meat consumption, may have been used as working animals due to their proclivity to cooperate with humans. There is evidence that pigs were used for threshing and planting grain in ancient Egypt (Masson, 2004, p. 36). In pre-industrial Britain, dogs and horses were also significant sources of power, and oxen were not eaten because of their importance in ploughing (Thomas, 1983). This reliance on animals as a source of labour power, however, came to a decisive end in Europe with the development of water, wind and steam power.

The rural pastoral still offered in much Western children's literature and film reflects a pre-Fordist model of the farm, such as might have been apparent in Britain and elsewhere in Europe, from the thirteenth to the nineteenth century, with most farms being relatively small and mixed and with a range of animals present, despite regional tendencies. This small-scale farming occurred on relatively sustainable pastures. The regionality of the rural landscape was apparent in the different kinds of husbandry and different kinds of products—the growing of different varieties of chickens, pigs, sheep and cattle, and the production of different sorts of animal products—such as the many local kinds of cheese that can still be found in Britain and France, for example. These reflected geographical conditions and constraints, with limited transportation possibilities and the perishable nature of animal foods. In Britain

> . . . this trend has continued up until the present day, with cattle and sheep production favoured in the wetter Western half of Britain, whilst the intensive pig and poultry units tend to be situated in the east, where the cereal crops used to feed them are grown. (Johnson, 1991, p. 33)

The seeds of the contemporary globalised animal food system, however, were to be found in this period and were tied to national interests and the domestic demands of the European dominant classes.

The process of colonization involved the development of an internationalised food system, which co-existed with the localised model in European regions. Extensive cattle ranching and sheep grazing on relatively unstable grasslands were the modus operandi of the farming system introduced by European colonization of the US, South America, Australia and Africa from the sixteenth to the ninteenth centuries. This system involved particular forms of exploitative social relations. On the one hand, there was the use of slave labour, displaced indigenous peoples and unwanted or exploited rural peasantries. On the other, landowning classes of sheep and cattle barons prospered, as did the exchequers of European nations through increased shipping wealth (Franklin, 1999, pp. 128–129).

As colonized territories became increasingly independent, and many drew in burgeoning immigrant populations, the ranching system, exploitative of both land and labour, became the model for an independent national system of production. The environmental impact of this system is well illustrated in the case of Mexico. Spanish conquistadors were followed by colonial pastoralists who assumed control of fertile agricultural land in the central highlands and began grazing sheep, shepherded by African slaves. By 1565, there were 2 million sheep in the region and by 1581, the indigenous Indians had been decimated by an epidemic imported by the settler community. Fields were turned into densely stocked pastures, which, by 1600, had been transformed into thorny desert (Cockburn, 1995, pp. 33–34). In the seventeenth century, the Spanish and Portuguese imported their native cattle into South and Central America (Velten, 2007, p. 28). This model was adopted in much of the southern US from the late eighteenth to the late nineteenth century, as US ranchers were seeking to increase profits by serving the expanding markets in Europe and importing cattle from Britain for this purpose. In many cases, diversity was replaced by species homogeneity in the process of increasing profitability. In South America, for example, Spanish colonialism established a ranching system around mission towns and villages in the seventeenth century. The sheep were descendants of the Iberian churro, which accompanied the conquistadors. Native American peoples developed and adapted this breed for suitability to rugged and harsh conditions and had relatively successful relations with these animals until the mass slaughter of the churro sheep by US federal forces in reaction to drought (Haraway, 2008, pp. 98–99). The hardy breeds were replaced by what were regarded as superior European breeds, and the churro have only very recently begun to make a comeback in initiatives around sustainability and support for traditional lifeways.

The demand of the English upper classes for fat-rich beef was an obsession throughout the first half of the nineteenth century, and the breeding methods pioneered in Britain were adopted elsewhere in Western Europe (Ritvo, 1990, pp. 45–50). Cattle breeding became an elite obsession and was represented in popular culture as a form of "patriotic duty" (Rogers, 2004, p. 15). Animals were bred to gargantuan sizes, and fat-rich beef was a quintessential sign of status. This was also expedient for the production industries—rendered cattle fat was itself a lucrative business (Velten, 2007, pp. 133–138). This demand, and the profits to be made from serving it, resulted in the 'cattelisation' of countries such as Argentina and Brazil and the replacement of species type in the United States. Jeremy Rifkin (1994, pp. 74–76) refers to this process as the "Great Bovine Switch," which saw the replacement of buffalo with cattle through the sponsoring of the hunting of buffalo, which led to their virtual and almost instantaneous elimination from the Western range lands after thousands of years of successful habitation. Thirty million buffalo were killed in around fifty years, and this opened the North American prairies to the cultivation of large numbers of cows. Initially, these were the classical longhorns of the brief 'cowboy' interlude. However, the development of 'homesteading' in the 1880s and the fencing of the open range enabled by the revolutionary new invention, barbed wire, led to the demise of the longhorns. Overgrazing and desertification, combined with fencing, kept longhorn from migrating to find food and avoid drought and severe winter weather resulted in the deaths of millions (Jordan, 1993, p. 80; Velten, 2007, pp. 149–150). The longhorns were replaced by British cattle breeds, such as Devon, Aberdeen Angus and Herefordshire (Velten, 2007, p. 150). This switch from buffalo to cattle was not only a colonialism of species but a strategic policy underpinning the forced resettlement of Native Americans on reservations (Hine & Faragher, 2000, p. 317).

The colonial model of meat production was further enabled by the development of refrigerated shipping, which made it possible to ship meat to Europe from the US, South America and Australia (Franklin, 1999, p. 130). Such ventures were particularly profitable in South America, primarily in Argentina in the eighteenth century and in Brazil in the nineteenth century (Velten, 2007, p. 153). In addition, meat processing plants were established in order to produce cheap meat products for working class consumption, such as the famous Liebig spread that was produced at the English-owned factory at Fray Bentos in Uruguay (Rifkin, 1994, p. 147). This enabled Europeans to consume greater quantities of meat, but in order to make the best use of the potential market in Europe, the price had to be minimised by intensifying production and saving labour costs through increased mechanisation, and it is to this that we now turn.

Industrialising Meat

Intense profitability was enhanced by the ability of manufacturers to extract 'products' from animal bodies. In writing of Chicago's famous Union Stockyards in the early years of the twentieth century, the novelist Upton Sinclair (1906/1982) described the way in which animal slaughter impacts everyday life, as the many 'lesser industries' that are maintained by the slaughterhouse profit from every part of the animal. Slaughter is but one element of the disassembly process in which animals are made into shoe polish, glue, soap, fertiliser and hairbrushes, in addition to fats, oils, meat and leather. William Cronon (1991) argues that the Chicago stockyards, which opened in 1865, were a crucial element in a complex network of technologies, agricultural practices and products. The railway network both facilitated and extended the transport of cattle to the yards and animal products out of them. Rail networks and the development of refrigerated carriages enabled connections among the productive elements of the meat industry (grain farmers, farmers of 'livestock') and the stockyards and their associated businesses. These innovations were also a key factor in overcoming traditionally seasonal patterns of supply.

The early meat factories of Chicago have been the model for production in the developed world. The social composition of the workforce is also little altered (apart from the use of child labour), with the continued use of migrant workers and those with few skills and other job prospects, in one of the most hazardous and poorly paid occupations (Marcus, 2005, p. 229; Nibert, 2002, pp. 66–69; Torres, 2007, p. 45). There are high levels of injury and death for workers in slaughterhouses and meat cutting plants and a cavalier attitude to both the health and safety of the workforce and to diseased meat. The journalist Charlie LeDuff wrote an article based on his observations at a pig slaughter and processing plant, which was published in *The New York Times* in 2000. He describes a deeply segregated place in which different communities (blacks, Mexicans, Indians and whites) all have different work stations, are segregated in different roles and practice self segregation in locker rooms and the cafeteria as well as the local bars away from the factory (LeDuff, 1999/2003, p. 184). The interviews of slaughterhouse workers conducted in the US by Gail Eisnitz made clear that slaughterhouses were not just places of fear, neglect and extreme cruelty endured by 'meat' animals but places in which human beings are brutalised (Eisnitz, 1997, p. 85). In the US, up to 100,000 cattle can be killed every 24 hours (Rifkin, 1994, p. 154). The pace of the slaughter line and conveyer belt meat cutting means that turnover of staff is high despite significant levels of local unemployment around 'meat plants.' The monotony is such that, "You hear people say 'They don't kill pigs in the plant, they kill people'" (LeDuff, 2000/2003, p. 185).

Similar conditions can be found in contemporary Britain. The overwhelming majority of the animals killed for food are killed in privately-owned slaughterhouses, and most butchering takes place in large packing factories that are constantly searching for labour through agencies and pay workers poorly. The work in both slaughterhouses and packing factories is physically arduous—moving stunned animals in order to shackle them, operating power saws, unloading frozen carcasses at an incredibly fast pace or seeing a carcass chopped, wrapped and boxed, all in 20 minutes. In the packing factories, operatives do not have any particular feelings about cutting up dead animals—as one put it, "We could be doing anything really. Well, anything really boring." The monotony of the labour is such that "every day lasts a lifetime." Boxing the cut meat is generally seen as the worst task: "It drives you mad. Literally. The 'freak show' that's what we call it. 'Cause they all look like freaks when they come out of there" (interviews, meat cutting operatives, London). Slaughterhouse and meat packing workers are poorly paid for long hours and for tedious, dirty, repetitive work using dangerous tools. They often work in excessively hot or cold temperatures and sustain injuries from animals, other workers and their own errors in a pressurized environment in which speed is of the essence.

In addition, fewer and fewer waste products are present, as increasingly, food can be 'reclaimed.' This has been dependent on the development of various new processes that enable the extraction of even greater profits from the bodies of animals. From the mid-nineteenth century, the meat industries of the US and Europe (in particular, Germany) began to use by-products from slaughterhouses, such as fertilizer, glue, buttons, combs, felt, margarine and glycerine (Nibert, 2002, p. 49). Today, the food industry has particularly benefitted from new chemical and mechanical interventions. For example, the filling of many processed meat foods involves 'mechanically reclaimed meat'—bone slurry, connective tissue and so on. The practice of 'reclaiming' meat has significantly contributed to industry profits, as waste is minimised, thereby reducing costs, and money is being made out of parts of animals that twenty years ago would have been discarded (interview, Smithfield Meat Market, London).

The profitability of processed meat products has been reliant on other developments in the technology of distribution, primarily the development of car culture and transport infrastructure. From the 1940s, the development of the road network in and between cities and the increased availability of the motorcar across the social spectrum enabled the development of fast food and its distribution at roadside restaurants. There is a particular geography of processed meat. The Mc-

Donald's corporation, from its very beginning, has analysed road networks and potential developments in citing its outlets (Ritzer, 2004, p. 219) and new sites are selected almost automatically with the use of geographic information systems (Schlosser, 2002, p. 66). The cycle of technological development and food innovation shapes food choices and fashions, as can be seen with the invention of the microwave oven and the development of 'ready meals,' often reliant on processed meat products—a fast food for the home (see Fine, Heasman, & Wright 1996, p. 206).

Technological developments do not only concern networks and outlets of distribution but the slaughter of animals itself. In the recent past, slaughter and butchery were closely linked. In Britain, before 1945, butchers usually had a slaughterroom 'out back,' and older men within the contemporary industry tend to see such 'old-fashioned' 'family' butchers as men of skill that form part of a romanticized past of the meat trade. Animals would be killed by being battered over the head with a pole axe, a hammer with a hook on the end (interview, slaughterhouse manager, Romford, Essex). Within the slaughter business, the technologies of killing are usually seen to have improved in terms of animal welfare. Yet the main changes in slaughter technologies were about maximising profits across different branches of the industry; they were not concerned with animal welfare. For example, the introduction of preslaughter stunning was primarily for the purpose of speeding the slaughter lines and improving meat quality (Burt, 2006, p. 127). Such standardization has little effect upon issues of animal welfare; the main concern of European regulations and directives are to eliminate bad practice in the area of food hygiene (Ministry of Agriculture, Fisheries and Food [MAFF], 1991).

Brave New Farm
By the 1920s, the US was leading the way in the mechanization of animal agriculture and millions of diversified small family farms had been replaced by specialist, large, corporate enterprises. Important in this transition was the development and use of tractors, replacing mules and horses in plowing and hauling. Technological innovation led to the development of a grain surplus in the US which, in turn, promoted the use of cheap grain by expanding meat producers (Nibert, 2002, pp. 102–103). Despite this, prior to the 1950s in Europe and America, most farms were family-owned or rented and family run, rather than corporate, and many farming practices, though larger in scale, remained similar to those deployed a century before. From the 1950s, one of the most important technological developments was the confinement of chickens for both eggs and meat; this was a means of significantly increased 'efficiency' and thus profit. Such farming maximizes land use through intensive housing and minimizes labour time as animals are in situ and fed automatically. The saving of labour costs has been dramatic.

In the US, one person may manage up to 150,000 laying hens (Mason & Finelli, 2006), although not all animals adapt to being permanently incarcerated:

> Particularly "advantaged" by these developments have been pig and poultry, especially chicken, due to the "high conversion efficiency of these species" . . . . The number of days taken to fatten a bird to 4lb declined from sixty to thirty-nine days between 1966 and 1991 and the amount of feed has fallen from 9lb to 7.75lb. (Fine, Heasman, & Wright, 1996, pp. 207–208)

Technology has been crucial in this process. The discovery, for example, that vitamin D supplements in chicken feed enabled animals to be housed without any access to natural light made indoor chicken-meat production a possibility (Mason & Finelli, 2006, p. 105). Whilst the bodies and minds of chickens endured intensely overcrowded, barren and polluted conditions, the postwar boom in the chicken business, particularly in the US, attracted the attention and investment of large pharmaceutical companies that developed treatments for diseases and 'unwanted' chicken behaviour.

Animal bodies themselves have been intensely modified to ensure suitability for industrial conditions and thereby enhance profit:

> In 1946, the Great Atlantic and Pacific Tea Company . . . launched the 'Chicken of Tomorrow' contest to find a strain of chicken that could produce a broad-breasted body at low feed-cost. Within a few years poultry breeders had developed the prototype for today's 'broiler'—a chicken raised for meat . . . . (Mason & Finelli, 2006, p. 106)

Following the successful intensification of chicken-meat and chicken-egg production, the 1960s saw the development of intensified and highly automated systems for growing other birds, pigs, cattle and sheep. Keys to success were automated feeding and watering systems, and for indoor-raised animals, the elimination of bedding and litter through the development of different kinds of food conveyance systems, cages, stalls, pens, forms of restraint and slatted floors over gutters or holding pits. Intensification has been applied to animals raised outdoors, and the cattle 'feedlot' of the US is the strongest example of this. Feedlots are fenced-in areas with a concrete feed trough along one side; they were developed in the context of depleting soil through overgrazing and surplus corn production, from the early years of the twentieth century. With nothing else to do, and stimulated by growth-promoting hormones, contemporary feedlot cattle eat grain corn and soya, which may be 'enhanced' with the addition of growth-promoting additives such as cardboard, chicken manure, industrial sewage, cement or plastic feed pellets (Rifkin, 1994, pp. 12–13). Slightly less barren and automated are the cattle 'stations' predominant in Australia and Central and South America. Here, cattle compounds are simply moved around when land becomes overgrazed (Velten, 2007).

The estimated global figures for animal killing are enormous. For example, 50 billion chickens and 1.3 billion pigs are slaughtered annually (Compassion in World Farming [CIWF], 2009, 2010). This scale of production has been enabled by the adoption of intensive farming methods: enormous profits are made from intensive farming in terms of the personal wealth of the owners of animal agriculture companies and their investors (Marcus, 2005, p. 5; see also, Torres, 2007, p. 45). On the other hand, the costs of animal products have remained relatively constant due to efficiency savings of scale and the 'improvements' in animal breeding that have enabled animals to be fattened to slaughter weight in almost half the time it took in the 1950s. The number of farms has thus been dramatically reduced. For example, in the US, the number of pig farms fell by more than two-thirds between 1992 and 2002 (Marcus, 2005, p. 9). Conditions of work in factory farms bear similarities to those in slaughterhouses and packing plants—extremes of temperature, occupational and infectious diseases, in addition to long hours and poor pay.

According to the US Department of Agriculture, only 2 percent of factory farms produce 40 percent of factory farmed 'meat' (Williams & DeMello, 2007, p. 21). Such enormous operations are part of the corporate giants of the US, such as Cargill, ConAgra, Smithfield and Tyson Foods, which are now 'vertically integrated' operations—that is, they own the breeding facilities, feedlots and indoor production units, slaughterhouses and packing facilities. Whilst production has increased and labour costs have been squeezed, soil and groundwater have also been damaged by the enormous monocrops for animal feed and by the hazardous amounts of waste generated by agricultural animals and the draining and contamination of irreplaceable groundwater stores (Gellatley, 1994, pp. 175–176). It is to this environmental and broader social impact that we now turn.

Food Colonialism, Intensive Production and the Environment

The previous section argued that the development of the global system of meat production was reflective of and contributed to forms of economic and social inequality generated by the intersection of capitalist and colonial relations. Despite this problematic past, however, intensive Western animal food production has asserted a positive developmental role in addressing present and predicted future food scarcity. Industrialised agriculture, including the production of 'food' animals, and the crops needed to 'grow' them, has been seen as a solution to food poverty. There are moves to 'democratise' diet, by encouraging Western intensive agriculture, particularly of 'livestock,' in regions of the South in particular. However, according to the Worldwatch Institute's "State of the World Report" (2004), citing UN Food and Agriculture Organization data, one of the most serious risks to the global environment is the expansion of intensive animal agriculture in Asia,

South America and the Caribbean. Industrialised animal agriculture is claimed to be a driving force behind all of the contemporary and pressing environmental problems that we face—deforestation, water scarcity, air and water pollution, climate change and loss of biodiversity (CIWF, 2002; Goldenberg, 2012), in addition to issues of social injustice.

New Colonialisms of Species

Projected population increases, combined with projected demands not just for food, but for meat-rich diets, will likely result in the decimation of remaining tropical and temperate forest, savannah and grassland in the Southern hemisphere by 2050 (World Bank, 2001). Such demand has led to corporate interventions, and the development of intensive animal agriculture in developing countries is currently proceeding apace.

Many US firms invested heavily in beef production in Central America in the 1970s and 1980s, and multinational corporations, such as Cargill and Ralston Purina, provided the technological support structure for the development of the Central American beef industry—from semen to grass seeds. Land reorganisation and the development of corporate farm enterprises, alongside the displacement of peasant populations, are "transforming an entire continent into grazing land to support the rich beef diets of wealthy Latin Americans, Europeans, Americans and Japanese" (Rifkin, 1994, p. 193). The most dramatic example is Brazil, whose government adopted a programme to convert the rainforest into commercially productive land in 1966, resulting in significant investment from US-based multinational companies in the Brazilian interior and the transformation of the Brazilian economy into the preeminent beef exporting nation. This is, in Rifkin's words, a new incarnation of "cattle colonialism" (1994, p. 199). Brazil and Mexico have devoted increasing amounts of their agricultural production to produce soy and sorghum to feed cattle, rather than corn to feed people, earning considerable export revenue as a result and contributing considerably to food insecurity (Gellatley, 1994, p. 154; Lappé & Collins, 1982, p. 11).

Robert Williams has argued that beef has contributed more economic and political instability in Latin America than any other export crop; for whilst sugar, coffee or bananas have clear and geographically bound limits, "cattle could be raised just about anywhere" (Williams, 1986, p. 158). This has created a new agricultural frontier in the region and politically empowered the cattle ranching elite, which, in states such as Guatemala in the 1970s and 1980s, were supported by repressive military governments inflicting displacement through extreme violence on indigenous peoples (Faber, 1993). In addition, the World Bank has estimated that, "since the 1960s, about 200 million hectares of tropical forest have been lost, mainly through conversion to cropland and ranches, the latter especially in Central and South America." (World Bank, 2001, p. 12). Whilst this region has been the most profoundly affected, rain forest has also been cleared in Southeast Asia for the growing of animal feed, such as the growing of tapioca in Thailand for sale to European Union countries. In Haiti, one of the poorest countries in the world, communities have been displaced to mountain slopes with poor soil, while much of the best agricultural land is used for growing alfalfa to feed cattle from Texas (Gellatley, 1994, pp. 152–160).

In addition, increased demand for cheap meat has led to the establishment of indoor production systems in poorer countries. In India, home to the greatest concentration of cows in the world, the population of 200 million cattle is still afforded the sacred protection in Hindu-dominated states, to the extent that killing cattle is regarded as a serious crime and the government maintains old-age homes for at least some of those too ill or old to roam the streets (Velten, 2007, p. 77, and extensively, Harris, 1987). Other species, however, have been more open to the Westernisation of farming practices. Battery systems for laying hens and the growing of chickens in broiler units are now widespread throughout the Indian subcontinent. Whilst these intensive methods have been promoted by agribusiness as a solution to current levels of malnutrition and hunger, the eggs and meat produced can only be afforded by social elites in poorer countries (MacDonald, 2010). The eating of meat and animal products is, in most parts of the world, seen as a form of desirable privilege and a mark of status and wealth. Such agricultural systems use huge amounts of scarce water, provide very few avenues of employment and make products largely exported to rich countries (such as the Gulf States).

The Politics of Meat and the Future of Animal-Derived Foods

In the aftermath of the Second World War, European states and the US set out to reduce malnutrition and hunger amongst their own populations with the promotion of cheap meat and other animal products. Rising levels of meat and dairy consumption became associated with social progress, as meat was not only an historic marker of status in the West but seen as necessary for good health. This was also promoted internationally by the UN, which, in the 1960s and 1970s, emphasised the necessity of increasing animal protein production and making such food increasingly available in poor countries (Rifkin, 1994, p. 131). It is difficult not to conclude that such initiatives were strongly influenced by Western governments driven by the interests of the multinational corporations based in their territories. Such initiatives ignore that pulses and grains have been the most common sources of protein across the globe, and that the ability of developing countries to feed their own populations successfully was significantly compromised by the replacement of staples such as corn, millet and rice, for monocultures to supply the livestock feed industry.

In the 1980s and much of the 1990s, the Common Agricultural Policy of the European Community/European Union also encouraged intensive animal farming through systems of grants and subsidies, which explicitly favoured the equipment and buildings of intensive production rather than improvements to land in which animals might be raised (Johnson, 1991, p. 181). More recently, however, the UN Food and Agriculture Organization report, *Livestock's Long Shadow*, concluded that animal agriculture is a greater contributor to global warming than the combined effects of all forms of transportation (Steinfeld et al., 2006). The deployment of Western agricultural models and the spread of Western food practices have significant implications for the environment in terms of the undermining of biodiversity, localised pollution, soil damage, rainforest depletion, and the contribution of 18 percent of all greenhouse gases. The technologies of animal agriculture have made meat production extremely profitable, resource hungry and wasteful. Considering the resources involved in breeding and growing a single beef cow, journalist Michael Pollan argues: "We have turned what was once a solar-powered ruminant into the very last thing we need: a fossil-fuel machine" (Pollan, 2003).

It may be that with apparent concern about climate change demonstrated by international organisations and the incontrovertible evidence of the role of animal farming in contributing to environmental hazards, national and international policy proclivities will shift. We have also seen increased public awareness in the West about issues of farm animal welfare. States, international organisations and even agribusiness corporations have deployed animal welfare arguments and combined them with ideas about meat quality in order to instigate moves, such as the banning of battery cages and sow tethers and gestation crates within the European Union. Yet, at the time of writing, two unsettling processes are at work.

First, the complex international system of animal agriculture seems set to expand, and concerns for animal welfare or environmental damage by livestock farming appear insignificant in the face of development driven by multinational corporations. Thus we see that the feedlot system is being exported to beef farming beyond the US, as is the practice of intensive dairy farming where cattle are kept permanently inside, in small stalls. Intensive dairy farming has been adopted in European Union countries, and at the time of writing, a number of 'mega-dairy farms' are being established in the UK. Key targets for Western-based agricultural corporations in the near future, however, are parts of the Indian subcontinent and Africa. Western intensive models, promoted by the agribusiness giants, are set to transform farming in some of the poorest countries in the world, just as they have transformed much of Central and South America in the latter twentieth century.

Second, a very small but notable development in particular Western countries (the Netherlands, Norway and Sweden) has been research and development activity around 'in vitro meat.' This involves the growth of muscle tissue in laboratories with the intention of developing it for consumption as food. This technology, essentially, is a form of stem cell science that transplants tissue engineering techniques from biomedicine into agriculture. Advocates for such technology such as the 'In Vitro Meat Consortium' or the campaigning group 'New Harvest' promote in vitro meat as a solution to the problems of animal cruelty and environmental damage caused by meat production by animal farming. In vitro meat is promoted as a social good, able to reduce pollution, deforestation and greenhouse gas emissions associated with livestock production methods (see Stephens, 2010). It is also held to have health benefits in being potentially 'fat-free' and free also of steroids and various problematic chemical additives. The absence of 'actual' animals also has meant animal rights campaigning organisations such as PETA (People for the Ethical Treatment of Animals) have seen in vitro meat as suitable for consumption, rather ironically, by vegetarians and vegans!

Yet in vitro meat cannot be seen as unproblematic. Even the economic assessments undertaken for the 'In Vitro Meat Consortium' suggest that certainly initially, this 'meat' will be intended for the high-end niche market in Western countries, priced above free range organic meat products (Stephens, 2010). As such, it seems an unlikely element of a solution to food poverty in developing countries. In addition, the production techniques suggest that in vitro meat production would place pressure on water resources, raising questions about the environmentalist credentials of this new technology. Both these future scenarios suggest that the production and consumption of meat remain firmly within research, development and policy scenarios, despite the increasing presence of critical voices.

The processes of exploitative relations that characterised the development of the international system of production and consumption of animal-based food continue, therefore, to play themselves out on new territories. Those territories are not only geographic but biological, as the bodies of animals are opened up to new forms of exploitation.

Conclusion

The food we eat is politically constituted. This chapter has mapped the economic and political trajectories of the development of the modern animal food industries. It has argued that developments in technology have been crucial to these processes, but both technological development and food production and consumption are developed through institutions and practices that are historically situated and socially produced. These are importantly shaped by systemic relations of social power—capitalism and colonialism—that are evident in the historical development of apparently modern and modernising societies. It has been argued that these systems of social relations continue to shape the trajectories of animal agriculture.

The production of animal-derived foods also has significant environmental legacies, from the development of the American prairies through the ranching of cattle (or 'hoofed locusts') in the nineteenth century, to the water hungry and water polluting factory farms that currently spread increasingly across the globe. The intensive production of stock is being adopted as a model for agricultural development in poor countries. The establishment of Western intensive production and the promotion of Western eating habits have increased social inequalities and will have disastrous consequences for both the security of human communities in some of the poorest parts of the world and the lives of the huge numbers of those nonhuman animals raised for human food. For the chicken peeking over the new horizon of animal food production and surveying the spreading use of battery cages and broiler production in South and Southeast Asia, the future looks grim. We might well share her unease—the sky could soon be falling on our heads.

This chapter has described and critiqued the historical processes and current constitution of the global animal food industries. Perhaps the most significant innovations here were the development of first, the technology of disassembly pioneered in Chicago, and second, the development of animal breeding techniques and methods of animal raising that have resulted in intensive production systems that are currently spreading apace across the globe. These processes have not only resulted in horrendous conditions for those species raised as 'meat,' environmental degradation and food poverty in poorer counties, but they have embedded inequalities, in terms of the poor conditions and bad pay of workers (in this case, in slaughterhouses and 'factory' farms). These conditions are reproduced both within sectors of the animal food industry and migrate from sector to sector both within and beyond food production, as discussed elsewhere in this book.

PART II

Technology, Development and Inequality

In this part of the book, we focus on issues concerning technology, development and inequality. This includes relationships between ICT use and development, which are considered in two chapters. Education, health services and food production, distribution and consumption play critical roles in development, so this section also includes chapters on these subjects. International organisations such as the World Trade Organisation and the World Bank try to persuade developing countries to invest heavily in ICTs and to open up their public services to international capital. This part of the book serves to cast serious doubt on whether such international organisations' policies actually benefit developing countries.

In Chapter 5, Miriam Mukasa considers the cultural implications of the consumption of ICTs for development. Organisations such as the World Bank try to persuade developing countries that they should invest heavily in ICTs as they are an engine of growth. Such propositions rely on the belief that ICTs can help developing countries narrow the gaps in productivity and output that separate them from industrialised countries—and even that they can 'leapfrog' stages of development into the information economy. However, most developing countries are unable to cope with the new technological paradigm or exploit its potential.

Peter Senker considers the social and geographical context of health service provision in Chapter 6, comparing regional differences in resources and evaluating the ability of systems to deliver and deploy health care technologies.

The problematic assumptions underpinning the use of ICTs for development are also taken up by Allyson Malatesta in Chapter 7, which focuses on the impact of ICTs in education. ICT producers have promoted their products as means through which excellent education can be made available to diverse and widespread communities. The World Trade Organisation believes that public services should be opened up to international capital and that this would benefit both globalization and education. However, critics maintain that further opening up the markets of poor nations to transnational corporations is liable to create greater inequalities between rich and poor nations.

In Chapter 8, Peter Senker questions the prevailing vision of a 'modern' agriculture from the Green Revolution to the current Gene Revolution as a standard, preferred pathway to development. Such a perspective centres on technology, production and growth. Key elements of the modern agri-food 'system' involve a wide array of external expensive inputs such as research and development, fertilizers, seeds and irrigation. Centralized technology-driven economic growth through sustained innovation and trade is envisaged as providing pathways out of agriculture or a shift of subsistence-oriented 'old' agriculture to a modern, commercial, 'new' form of agriculture.

FIVE

The Cultural Implications of the Consumption of ICTs for Development

Miriam Mukasa

This chapter considers the current debates in ICT (information and communication technology) for development discourse and the cultural implications of the consumption of ICTs for development. ICTs have gained prominence in organisational and development thinking. This prominence is evident in the numerous policies and initiatives worldwide aimed at accelerating the adoption of ICTs for development. The rationale for adoption of ICTs is usually presented in terms of bridging the digital divide, enabling the millennium development goals, and enhancing socioeconomic development. The perspectives that dominate the debates, however, reflect rationality based on modernisation and fail to consider the role ICTs play in reinforcing inequality.

Drawing on the biography of artefacts framework (Pollock & Williams, 2008), and on observations made from a four-year study of adoption, embedding and use of commercial off-the-shelf software (COTS) in a large higher education institution (HEI) of a least developed country (LDC) in Sub-Saharan Africa, this chapter traces the history of COTS and its production/design to demonstrate how the production and consumption of ICTs reinforce inequality and dispels the myth that ICTs can lead to development. The production and consumption of ICTs are inherently political and have inequalities of all sorts, making current notions that ICTs can lead to development questionable.

The case study institution is one of the oldest and most prestigious centres of learning in Africa. The institution clearly recognised the importance of integrating ICT in all its functions as a critical element of its strategic plan. As such, adopting COTS was seen as a way to meet the institution's objectives. The institution has also been one of the HEIs at the forefront of ICT adoption in Sub-Saharan Africa and has been a recipient of donor funding for ICT projects from various donor agencies, including the World Bank.

ICTs for Development

The debates around ICTs for development are centred on the positive role ICTs can play in socioeconomic development and the opportunities available for developing countries in the use of ICTs. One side of the debate tends to focus on improving organisational efficiency and productivity through implementing information and communication technologies to facilitate 'best practice' and enhance development, while the other emphasises the importance of information as key to the developmental process and the ability of ICTs to process, store and transmit information to facilitate decision-making processes and enhance knowledge (Ducombe & Heeks, 2002; Mansell & Wehn, 1998; Van Der Velden, 2002; World Bank, 2001, 2002). Emphasis is also put on the potential to use information effectively as a key resource in creating an 'information society.'

The overall idea is that ICTs are an engine of growth, and if these countries invest heavily in ICTs, they will be modernised. The World Bank, for example, argues that, "this new technology greatly facilitates the acquisition and absorption of knowledge, offering developing countries unprecedented opportunities to enhance educational systems, improve policy formation and execution, and widen the range of opportunities for business" (World Bank, 1998, p. 9). Following this reasoning has been the belief that gaps in productivity and output that separate industrialised and developing countries (Steinmueller, 2001) can be narrowed, enabling leapfrog into the information economy (Schech, 2002). Suggestions have even been made that the use of ICTs can enable developing countries to leapfrog stages of development (Davison, Vogel, Harris, & Jones, 2000; Hudson, 2001) and select new positions on the technology curve (Flemming, 2003). Yet despite the prevalence of such beliefs, there hasn't been any evidence to support the link between ICTs and development. Even where some developing countries such as China have made inroads in terms of economic development, this cannot be attributed to ICTs. Arguably, a combination of factors, including low production costs enabled by the abundance of cheap labour and low cost capital and other forms of capabilities such as education, have contributed to this progress. Such are the myths surrounding ICTs and as noted by Burnett, Senker, and Walker (2009, pp. 1–15), myths that have to be constantly promoted for capitalism to be legitimised and sustained.

One thing to note, however, is that there is no established convention for the designation of a 'developed' or 'developing' country (United Nations, 2010). This doesn't only raise concerns about the measurement of development but the concept of development too. For example, The World Bank classifies developing countries as low- and medium-income countries based on their gross national income (GNI) per capita (previously gross national product [GNP]). As such, classification is purely economic and limited, given the other equally important aspects of development such as freedoms (see, e.g., Sen, 1999). The World Bank notes, however, that, "this does not imply that all countries in the group are experiencing similar development or that other economies have reached a preferred final stage of development" (World Bank, 2011). As an example, 49 of these countries are currently designated as LDCs, including the case study country, and have remained marginal in the world economy, owing to their structural weaknesses and the form of their integration into the world economy (United Nations, 2010). A recent classification of LDCs by the UN, on the other hand, is based on three criteria: 'low income,' 'human asset weakness' and 'economic vulnerability' (United Nations, 2011), all of which are reflective of the lower status of developing countries as compared to developed countries and the inequality that exists between and among countries of the world.

While the importance of ICTs for development cannot be underestimated, there are few signs that enthusiasts for investments in ICTs consider these factors or understand the role ICTs play in reinforcing these existing inequalities. For instance, technological leapfrogging, which is cited as a key benefit ICTs can provide to developing countries, is mythical in that it runs in contradiction to the cumulative nature of the learning processes (Hobday, 1994), and an unprepared economy or society, as is the case with most developing countries, would be unlikely either to cope with the new technological paradigm or to exploit its potential. Empirical evidence also suggests that ICTs begin to deliver GDP per capita growth only after a 'certain threshold' of ICT development has been reached (McCauley, 2004, p. 9), a threshold which, arguably, cannot be reached by most developing countries given that they are normally consumers rather than producers of ICTs and given the many challenges these countries face. This is not to say that there are no developing countries that have benefited from ICTs. South Korea and Singapore are, for example, cases where economic growth has been achieved as a consequence of involvement with export-oriented production of ICT, in particular, consumer electronics (Senker, 2000), with subsequent changes in status from 'developing' to 'developed' countries.

Nonetheless, the process of production and consumption of ICTs is a complicated one, and it is marked with inequalities, uncertainties and byways that technological innovations configure their users (Silverstone & Mansell, 1996), ways that clearly benefit some and not others.

In fact, it is increasingly evident that the proposed benefits of ICTs are not being experienced by everyone as reflected by the emergence of a significant digital divide. Walsham's (2010) analysis of the literature on the contribution of ICTs to the development goals of India also substantiates this and is reflective of how ICTs reinforce inequality. The analysis reveals that ICTs have contributed to 'economic facilities' through various initiatives such as e-government services, telecentres and ICT-facilitated agricultural supply chains and computerised land reform. However, it is those already in relatively privileged positions who tend to benefit from these initiatives, as opposed to the very poor (Walsham, 2010, p. 16). Additionally, the analysis notes that the poorer states of India, such as Bihar and Orissa, are excluded from the ICT-based literature, in contrast to the reported work in relatively richer states, such as Kerala, Andhra Pradesh or Gujarat (Walsham, 2010, p. 16). As such, when ICTs are presented as the magic bullet for the woes of the poor, the competing political aims, social values and competing theories of social change seem to be ignored. In particular, ways in which ICT production and consumption reinforce existing social and economic inequalities are overlooked.

Consumption, Culture and ICT

According to Campbell (1995), consumption can be defined as the selection, purchase, use maintenance, repair and disposal of any product. As such, social and cultural approaches to consumption tend to frame consumption as an economic process that is concerned with the consumption of goods and services (Evans & Jackson, 2008, p. 6). ICTs are also goods and services and their consumption is also framed within this context. However, such a conceptualisation is problematic because it takes a narrow view of the consumption process of ICTs. Consumption, as Fine (2002) argues, cannot be understood purely in economic terms. Consumption is structured by uneven distribution of economic and cultural forms of capital throughout society (Carlisle, Hanlon, & Hannah, 2007). Hence, the analysis of consumption should take consideration of the capitalist nature of the economy and the different capacities to consume of the different classes (Fine, 2002), and the same should be the case for analysis of the consumption of ICTs for development.

Take, for example, information, which is central to ICTs. Consumption of ICTs involves exchange of information from those that have it and those that do not. As such, information is viewed objectively as a valuable commodity, which can be moved from one context to another. But given the economic context within which ICTs are framed, those that have or own the information cannot fully reveal their knowledge without destroying the basis of trade. This creates a problem of asymmetric information, whereby consumers cannot fully determine the value of the information before buying it (Hoeckman, Maskus, & Saggi, 2004, p. 4). This means they cannot use the information to their benefit. This is of significance given that ICTs are advocated as technologies that can bring benefits to all. It also reflects the broader challenges associated with the relationship between power, culture and knowledge.

Additionally, in an international context, information problems are more severe and the enforcement of contracts is difficult, something that can arguably be attributed to the structure, concentration and ownership of the industry and associated power relations. As an example, for the case study institution, there were limitations in terms of the information embedded in the ICT as well as in terms of the information communicated about the access and use of the ICT. Members of the project team also noted that, two years down the road, the institution had not renewed the support contract; what they had was only a component of the system. It was a system that wasn't fully functional and was not meeting the requirements of the institution. Arguably, an ICT contract is a commodity, as the ICT it serves and the content and way it is consumed are critical to successful adoption of ICTs. Without the right level of skills to negotiate contracts, and facilities to enforce them, ICT contracts exacerbate the digital divide.

The impetus to use information in new ways is also driven by competitive pressures of international business, and the existing inequalities and values that created them and are deeply entrenched are completely ignored; the digital divide, which has become one of the most cited woes of uneven development since the onset of informational capitalism (Parayil, 2005), is reflective of this. Most of its analyses are presented in a narrowly construed instrumental sense of the absence of a technical artefact for a large number of people in the world that deprives them the ability to access ICTs, and the only solution to this is to provide access to ICTs. But as Parayil (2005) argues, the digital divide is not an accessibility issue, as it is made out to be. Neither is it a technological problem due to the absence of ICTs. The digital divide is an equity issue. It is the outcome of wider inequalities embedded in society. To put it simply, ICTs just reinforce the digital divide as reflected by the world ICT indicators and the many ICT initiatives and projects that continue to fail in developing countries (Avgerou & Walsham, 2000; Heeks, 2002) and continuously fail to meet intended objectives. As an example, while the number of internet users doubled between 2005 and 2010, the percentage of the internet user population in the developing world was far behind that of the developed world. In the case of Africa, for example, the continent was way behind the world average of 30 percent and the developing country average of 21 percent (ITU, 2010).

A number of studies have attributed these disparities to contextual problems associated with differences among models of developed countries' design and applications in a developing countries context, what Heeks (2002) calls the 'design-reality gap' and to different value systems and different concepts of rationality (Avgerou, 2002). Failures are also said to arise from strategic problems that relate to local, regional and national policy initiatives that influence the regulation and adoption of ICTs and operational problems, which are associated with economic, technical and skill deficiencies (Heeks, 2002). For instance, having a sound infrastructure is a prerequisite for adequate development and technology policies have to be integral to the promotion and management of sustainability of ICTs, aspects that are lacking in most developing country environments.

Misund and Hoiberg (2003) define technological sustainability as the ability for a technology to exist for a long time without major changes in hardware or software affecting its availability or durability. This includes operational simplicity, flexibility, maintainability, robustness and in case of organisations, capability of technical and managerial staff (Kiggundu, 1989). However, most developing countries have limited technological infrastructures and are faced with higher technological costs as compared to developed countries (Kiraka & Manning, 2002). Subsequently, they have to rely on donor funding for their ICT imports, funding that is often provided for a limited period (Harris, Kumar, & Balaji, 2003; Hudson, 2001; Kumar & Best, 2006) with implications for sustainability not only in terms of maintenance costs but also other recurrent costs associated with the ICTs. Indeed, this was the case with the ICT project for the LDC case study institution whose experience of consuming ICT, despite having some positives, was extremely challenging.

For instance, at the time of adopting the technology, the institution had a handful of computers at its disposal and had to rely on donor funding to acquire more computers. Dependence on donor funding for ICT consumption also implies that there are limitations in terms of choice for developing countries, resulting in wider implications for development, as consumers are likely to be influenced to deploy fashionable ICTs without considering critical issues such as user capabilities and cultural feasibility. This is compounded by the fact that most of the ICT projects are characteristic of narrow interventions, pilot project orientation and inadequate focus on human resource development (Baark & Heeks, 1999; Braa, Monteiro, & Sahay, 2004). With regard to the case study institution, for example, the majority of staff couldn't even use a keyboard, which necessitated large-scale training that unfortunately couldn't be provided effectively due to a variety of constraints, such as finances, lack of policy coherence and skill constraints. Additionally, most of the users felt threatened by the new technology and were reluctant to engage with it, with some claiming that the terminology used didn't match what was used in their work, signifying the 'design-reality gap' that is characteristic of ICTs.

This meant that a few individuals could have access to the information system, thereby creating inequity among the different staff within the institution and failure to realise the objectives of the institution fully. Despite this, investments continue unabated, albeit at high costs, often without examining the question of why the ICTs are being deployed.

In this context, while there has been growing propensity to consume ICTs, analyses of ICT initiatives and programmes pay little attention to how the production of ICTs and the social meanings created within consumer culture possess symbolic force that can add to wider inequalities. ICTs enter our lives with meaning and the need for uses and the value attached to the ICTs is developed and redeveloped in particular settings. This process of consumption involves creativity, reappropriation and recontextualisation (McLoughlin, Rosen, Skinner, & Webster, 1999). However, the process is embedded in an environment of inequality that is being replicated the world over, and which, as Pieterse (2002) notes, evokes what has been termed the 'second great transformation,' the transformation from national market capitalism to global capitalism (Pieterse, 2002, p. 1024). This point is significant because ICTs are clearly linked with economic rationality, and we need a reminder about the ideological nature of rationality. We also need to understand that the meaning and effectiveness of an ICT can vary substantially across cultures. But today's ICT initiatives continue to be predicated on the assumption that everyone benefits from this technology, which is not the case.

The Politics of ICT Production and Consumption

Social studies of technology argue that the construction of sociotechnologies and the ensembles that emerge out of the process are a site of social struggles for status and control. As such, the technical and social are blurred in the real work of technology construction. Additionally, once the technology moves beyond its engineering origins, the black box comes to represent something that can be of value but whose origins and workings are not understood (Winner, 1993, p. 431). This means that once a technology has been successfully black boxed, it is inscribed with particular meanings, uses and assumptions that constrain and to some extent shape the user (Akrich, 1992, pp. 205–224). Winner (1986) demonstrates this by highlighting the case of the architect, Robert Moses, who excluded from his showcase development poor people and minorities who relied on public transport by ensuring that a bus driver wishing to reach Jones Beach in New York would not be able to pass beneath the low-hanging overpasses that cross the entry roads. Similarly, ICTs exclude the poor and the minority who don't have the ability to access and use the ICTs or who lack capacity to absorb this technology. For example, certain assumptions about users and society are inbuilt into the technology and these assumptions create constraints on the ways in which users can engage with the technology. As an example, the assumption that the predominant language of ICTs is that of English and is accessible to all societies is a significant constraint to the way users engage with ICTs, as is the assumption that everyone has the financial means or the knowledge and skills required to access and use ICTs.

Such constraints imply that technology is culturally situated. In fact, as Bjorkman (2005) argues, software is said to be tightly interwoven with culture. It is created by culture and it creates cultures. The implication here is that consumption is not only a symbolic activity through which people consume cultural meanings; it is also an activity through which productive consumption takes place (Fine, 2002), the impact of which goes far beyond the simple aspect of accessibility and use of ICTs. For instance, Boyle (2002) provides examples of ICT failures in the public sector and the general complexity of social change involved in ICT deployments in developing countries. He discusses the implicit Western assumptions inscribed into ICTs, which often mismatch cultural realities of developing countries and which impact on the users' capacity to adjust to and understand the new system in which they are involved.

This resonates with Feenburg's (1999, 2002) argument that technological artefacts are not products of pure engineering and design processes but are also the result of often-conflicting views of the world. Winograd and Flores also support this. They argue that, "things that make up software, interfaces and user interactions are clear examples of entities whose existence and properties are generated in the language and commitment of those who build and discuss them" (Winograd & Flores, 1986, p. 69). As such, ICTs structurally integrate communities into wider uneven networks of power (Thompson, 2004). The basic premise here is that the assumption of rational choice for the consumer is complicated by a diversity of interests, preferences, values and distribution of resources, all of which involve conflict, cooperation and negotiation. This alters significantly the notion that ICTs can be of benefit to all. Additionally, while ICTs offer exciting possibilities, enthusiasts of ICTs rarely face the issue that ICT is a commodity and that the consumption of this commodity is inherently political. What this implies is that, the paradigmatic premises of ICT and modernisation take a neutral view of ICT, which ignores the fact that it can be a mechanism of control if existing socioeconomic order prevails and can increase inequality and dependence of the underdeveloped or the poor on developed/donor countries. In this context, ICTs are taken for granted, limiting the ability to understand the cultural implications of their consumption, particularly for development.

Furthermore, when one considers the fact that the importance of ICTs is inevitably, largely constructed at a considerable remove in time and geographical and social space from their use, it is critical that ICTs are examined not solely in terms of the place where the user encounters the technology but also at the point of technology supply and the broader processes associated with the historical setting of the unfolding of the technology or its production. Most accounts of ICTs also ignore the extent to which consumption of ICT is intertwined with its production. Understanding the nature of this relationship and the associated processes can contribute to our understanding of the relationship between ICT and development, as the following example of production and consumption of COTS demonstrates.

The Case of COTS

COTS are standardized software packages that have become widely accepted worldwide. Examples of COTS include Enterprise Resource Planning (ERP), Customer Relationship Management (CRM), and other financial and administrative integrated systems. COTS are seen as the success story of the IT industry. The software is "designed to integrate the flow of information throughout the organisation while automating core activities of the organisation such as human resources, manufacturing and finance" (Gibson, Holland, & Light, 1999, p.1). The software is also said to capture 'best practices' and state-of-the-art processes of a vertical industry or business function (Hurwitz, 1998; Mabert, Sonni, & Venkataramanan, 2003) and is marketed as offering 'strategic advantage' to or-ganisations by giving them more control and flexibility. As such, COTS is seen as a leading edge technology that can facilitate organisations in developing countries to effectively respond to developmental opportunities while simultaneously devel-oping efficient administrative processes that are critical for national development.

Recently, a number of organisations in developing countries, especially those with strong linkages to multinational corporations, have already implemented or are in the process of implementing COTS (Molla & Bhalla, 2006) for their enterprise systems. Evidence suggests, however, that COTS infrastructures are the most expensive and the most difficult to implement (Sia & Soh, 2007). In this context, one wonders how the poor with limited financial resources can benefit from the technology to the extent of achieving their developmental goals. In fact, empirical evidence suggests that these projects are failing in developing countries (Hawari & Heeks, 2010; Kamhawi, 2008; Rajapakse & Seddon, 2005). As such, one cannot help question the efficacy of these ICTs for the poor if they can't access and use them to their benefit.

The difficulties of adoption are said to arise from problems associated with particular features of the software implementation, such as customization (Dav-enport, 1998; Hurwitz, 1998; Sow, Kien, & Yap, 2000). For example, Kumar and Hillegersberg (2000) and Sow et al. (2000) highlight problems of mismatches between the underlying data and the process models and the organisation struc-ture, which lead to considerable implementation and adaptation issues. For con-sumers, therefore, the downsides stem from risks inherent in the software not adequately matching their requirements or their 'context,' a factor confirmed by an ICT manager of the developing country institution who succinctly stated that, "There are implementations that can't go ahead because of the way we work/do things." The production/design of COTS and its consumption is also marked by a number of key but shifting moments, which arguably reinforce inequality and ultimately endanger development.

COTS Production and the Reinforcement of Inequality

Traditionally, software production involved suppliers developing close ties with customers, the "conventional wisdom being that increased knowledge of the user would lead to better design" (Pollock, Procter, & Williams, 2003). In contrast, with COTS or generic software, production is opaque to consumers because of fear of the software being identified with specific user organisations and not marketed widely. Programmers also work without concrete notions of users in mind (Pollock, Procter, & Williams, 2003). So, the final product is the important thing. As such, the software is commodified by means of a 'selective' and managed approach to user requirements (Pollock & Williams, 2008), whereby pilot organisations are selected and their specific features designed in the software. The idea is to produce software that embodies characteristics that are common to many consumers. This means, there is a shift from production for a few isolated users to production for a market. However, markets, as we know them, occur in the context of capitalism and are embedded in a particular political and social context.

For instance, in producing the software, the search for the pilots is based on criteria, which excludes 'others' in a variety of ways. Take, for example, the criterion for selection of participating organisations. This is based on the 'generic potential' of their needs and who they are a surrogate for (Pollock, Procter, & Williams, 2003). As such, suppliers decide which organisational practices will be catered to and which will not. Naturally, this has implications for the poorer countries, as their organisational practices are unlikely to be catered to. It also implies that organisations in developing countries end up having no control, as their specific features are excluded in the design of the ICT system. Given this scenario, developing country organisations end up having no ownership whatsoever of the technology, a factor that limits their ability to adopt and adapt the ICTs to their advantage.

The 'generification' strategy is also used to increasingly develop novel ways to manage the continued recycling and extension of the software (Pollock, Williams, & D'Adderio, 2007) and to manage the relationships between the vendors and the consumers, strategies that clearly favour the vendors and reinforce inequality, as demonstrated by the consumption of COTS. For instance, the software is designed in such a way that users have to upgrade their systems from time to time. The implication is that the poor cannot engage with the ICTs on par with the rest of the world. For instance, with regard to the case study institution, there were problems with upgrades, as indicated by the following statement from a database manager:

... We run the lower version of oracle and oracle is no longer supporting that version but upgrading to a higher version means you have to transfer data and we don't have the sufficient skills. This also translates into cost. And do we have that money?

Additionally, you have technical support being provided at a distance and mediated through technology. As such, it is not based on local and situated frameworks. Instead, you have a globalized face to portal form of support, which generates problems for consumers that lead to delays in responding to problems, and in some instances, no one taking responsibility for solving these problems. For instance, adoption of COTS requires local presence of intermediaries in developing countries. However, these intermediaries serve their own interests, as highlighted by the case study institution. An ICT manager from the institution noted, for example, that the institution cannot communicate with the vendor directly, which makes it very difficult to solve the problems that are encountered. He went on to state the following: "Generally it takes 2–3 days which is outrageous . . . . If we had the programmer come in and directly sit here for a month, the problem would be solved but the business man has to put his cost . . . ." Subsequently, production and consumption of the technology become a continuous process of construction and reconstruction, which reflects endless dynamic interplay between local and global dimensions.

In view of this and given the wider industry dynamics that surround software and how the market of software packages is constituted, ICTs cannot bring the same returns to developing countries as has been done to developed countries. The consumption of COTS clearly demonstrates this. It shows that consumption of software does not allow for local ownership, as it is alien to the consumers, with implications for equity. In this context, the vision of transformation, modernisation and subsequent development becomes an illusion.

Implications for Development

Development is said to be a founding belief of the modern world (Peet & Hartwick, 1999). It is said to be an idea, an objective and an activity, undertaken to accelerate economic prosperity and social well-being, involving a shift away from conditions of life that are perceived as unsatisfactory, toward those that are significantly better (Kothari & Minogue, 2002). In this context, development entails human emancipation, which, according to Peet and Hartwick (1999), is by way of liberation from the vicissitudes of nature through advanced technology and self-control over social relations and conscious control over the conditions under which human nature is formed. As such, "development entails economic, social and cultural progress including finer ethical ideals and higher moral values" (Peet & Hartwick, 1999, p. 1).

However, development is still elusive and there are ongoing debates about what it constitutes (Piertse, 2002) and criticisms of its failure in terms of development programmes and initiatives to live to their expectations. Development literature also indicates sustained critiques and debates about the dominance of one particular ideology (see, e.g., Escobar, 1995; Kiely, 2007; Mehmet, 1999), the exclusion of certain groups of people from the development project (De Rivero, 2001; Hammer, Bennet, & Wiseman 2003; Seligson & Passe-Smith, 2008), and the processes and procedures of development (Kothari & Minogue, 2002). As such, development is a complex phenomenon, "one reflective of human aspirations and yet, exactly because great ideas form the basis of power is subject to the most intense manipulation and liable to be used for purposes that reverse its original intent" (Peet & Hartwick, 1999, p. 2).

Measurement of a country's development is by way of the Human Development Index (HDI). The index is based on three tenets: life expectancy, knowledge and standard of living (United Nations, 1990), all of which contribute to the economic success of a nation. However, "measuring a country's development by levels of consumption of goods and services confuses the role of commodities by regarding them as ends in themselves rather than a means to an end" (Todaro & Smith, 2003, p. 19), a factor that is also echoed by the drive behind the ICT and development discourse.

Development is seen by some as something that can have meaning to people only if it is built on the way in which they see themselves. For instance, Sen (1999) sees achievement of development as being "thoroughly dependent on the free agency of people" (Sen, 1999, p. 4). As such development can only occur if freedoms that people have are enhanced or if people can have freedom of choice in the social, economic and political sphere. With regard to commodities, Sen's argument is that, what people do with the commodities of given characteristics that they come to possess or control, that is, their 'functionings,' is also equally important in our assessment of development. For Sen, functions generate the outcome and the capabilities facilitate the freedoms one can enjoy or exercise. The implication here is that ICT goods and services are important only if people can generate capabilities from them. In other words, ICTs can only be useful in light of their contribution to the consumer or user's capability set. This suggests that development should be about learning and questioning assumptions and this process is influenced by personal, social and environmental aspects.

However, the conceptualisation of ICTs in development and its consumption is strongly associated with images of inevitable progress through technology and abstract notions of instrumental rationality (McLoughlin, Rosen, Skinner, & Webster, 1999), values that influence specific processes of innovation as demonstrated by the case of COTS, values that imply that ICTs operate within the limits set by the nature of liberal capitalism as a global system and the practical and normative implications of its relationship with other forms of social organisation with which it coexits (Brett, 2000). Given this context, ICTs are aligned with a market economy, which, as we know, is characteristic of deep income inequalities, uneven development and asymmetries in intraregional economic growth in developing countries (Parayil, 2005). There doesn't seem to be any prospect of this changing any time soon. This free market also governs the major institutions of development and their policies. Take, for example, The World Bank's policy and its agenda. The agenda stresses competition and private partnership and is clearly market oriented, as demonstrated by its 1991 report, "The Challenge of Development" (World Bank, 1991). The report starts with a plea for a free market, and its partnerships with the International Finance Corporation (IFC) and the Multilateral Investment Guarantee Agency (MIGA; Urey, 1995, pp. 116–119; World Bank, 1994, pp. 35–40) are underpinned by support for private investors whose interests may not necessarily be conducive for development.

More so, under capitalism, the market process rewards those who own and control capital in any form. As such, Western interests typically enjoy monopoly and oligopoly powers and profit from trade in technology transfers intended to extend the capital base in developing countries, regardless of how appropriate such transfers may be (Mehmet, 1999). Value is created through intellectual property (IP) rights protection, and this automatically benefits the owners of the ICTs. The critical problem facing developing countries is that ICT stakes are higher than any communication battle in memory, though, according to Wu (2010), the form of battle is a familiar one. For those with no ownership and control of this technology and no capability and incentive to innovate, it is not clear how consumption of ICTs can benefit them to the extent of achieving their development goals.

Given this context, there are too many variables associated with ICTs that exacerbate inequality. These include interests and ideologies of various actors that are of significance with regard to the neutrality of ICTs. As such, ICTs are far from neutral in relation to development and, with such variables, for many, achieving development may be an unrealistic expectation, even in the distant future.

Conclusions

While economic factors may determine the outcome of ICT consumption, the process itself cannot be explained in economic terms alone. The production and consumption of ICTs involve a lot more than just adopting and using the technology. This suggests that one has to consider the historical, political, social and psychological factors that may contribute to the outcome. One also has to consider the relationship between ICTs, markets and institutions. The underlying idea here is that ICTs come with not only cultural inscriptions and meanings but they also embody politics. The examples I have presented and responses from the case study institution demonstrate these factors and confirm the fact that the consumption of ICTs is a complex phenomenon that is influenced by personal, social and environmental aspects. Given this context and the discussion regarding the production and consumption of ICTs, plus the challenges developing countries face, it is important to point out that the inequities that exist in the development project cannot be simply eradicated by ICTs. Instead, these inequalities are just entrenched by ICTs.

Rather than the acclaimed positives of ICTs, therefore, and given the importance these technologies are accorded with regard to changing the lives of the poor, it is evident that the production and cultural consumption of ICTs have wider implications for development, particularly with regard to equity. This is even more so given the internationalised supply of ICTs and consumption across different sectors, the different dynamics of today's ICTs as opposed to traditional bespoke software systems and the external nature of development that is based on the model of the industrialised world. It is these dynamics that help perpetuate inequality, and it is these dynamics that need to be examined carefully if the production and consumption of ICTs are to have a positive impact on development.

SIX

Health Care Systems, Technology and Inequality

Peter Senker

The health services a country needs and can afford depend on many factors, including the health problems suffered by its population, its climate, its geographical location and its wealth. This chapter presents a hypothesis that an effective way of improving a nation's health is likely to be by means of an integrated national health service, in which a range of health services, including both preventive and curative services, are delivered to individuals generally free at the point of use, and in which all the facilities delivering these services are publicly owned. All national health care systems include blends of different elements, for example, private insurance and private and public service delivery, so generalisations such as those in this brief chapter have to be treated with extreme caution.

Health care systems cannot be considered in isolation. The infrastructure—including availability of clean water, sanitation and adequate nutrition—is a crucial contributor to health. In developed regions, resources such as clean water and refrigeration, which are essential for conventional storage and delivery of some vital health supplies such as drugs and injections, are readily available. In contrast, in many areas of developing countries, refrigeration, clean water and other resources necessary for safeguarding human health are scarce. There are important, complex relationships between the health of the population and income inequality, and these are discussed in this chapter.

Technology—in particular biotechnology—has great potential for treating disease. But effective health service delivery systems are always essential, even though specific requirements vary considerably. For example, health education and the distribution and use of condoms are essential ingredients in the treatment of the control of HIV/AIDS. Improvements in sanitation are essential ingredients in reducing the incidence of water-borne diseases. Basic nutritional education can help prevent nutrient deficiencies (Daar et al., 2002).

The main causes of illness and death in developed countries are cancer and diseases of respiratory, cardiovascular and nervous systems. Communicable diseases are the main problem in the developing world: principal causes of death are respiratory infections, HIV/AIDS, infections at birth, diarrheal disease and tropical diseases such as malaria. Nevertheless, non-communicable diseases largely attributable to changing lifestyles, such as the increasing consumption of highly processed foods, are now rapidly increasing as causes of death and disease in developing countries (Lawrence, 2011).

There is still an inbuilt tendency arising from market forces that ensures that pharmaceutical development continues to concentrate on treating the diseases of the relatively affluent (Senker, 2000). Worldwide scientific research and technological development have been substantial, but resulting improvements in the health of the world's population have been far less than could have been achieved with better direction and distribution of the world's scientific and technological resources.

Health and Inequality in Developed Countries

Aneurin Bevan's (1952) perspective, vision and central arguments in favour of a national health service, free at the point of delivery, are as relevant today as when he wrote about them more than 50 years ago. Bevan was minister of health in the UK Labour Government, elected in 1945 after the Second World War, and he was primarily responsible for initiating Britain's National Health Service (NHS). Bevan wrote the following: "Modern communities have been made tolerable . . . by the activities of the sanitary inspector and the medical officer of health" (p. 73). In "backward countries" and

> . . . the backward parts of even the most advanced countries . . . the small well-to do classes furnish themselves with some of the machinery of good sanitation such as a piped water supply from their own wells and modern drainage and cesspools. Having satisfied their own needs, they fight strenuously against finding the money to pay for a good general system that would make the same convenience available to everyone else. The more advanced the country, the more its citizens insist on a pure water supply, on laws against careless methods of preparing and handling food, and against the making and advertising of harmful drugs. Powerful vested interests with profits at stake compel the public authorities to fight a sustained battle against the assumption that the pursuit of individual profit is the best way to serve the general good. (Bevan, 1952, pp. 72–75)

Bevan maintained that curative medicine should be driven by the same principles. He rejected both group insurance and the attachment of medical benefits to the terms of employment. He suggested that relating medical benefits to employment favours the strong over the weak—as it does in the United States today—and that insurance schemes are unnecessarily complex and bureaucratic (pp. 77–85).

> Society becomes more wholesome, more serene and spiritually healthier, if it knows that its citizens have at the back of their consciousness the knowledge that not only themselves, but all their fellows, have access when ill to the best that medical skill can provide. (Bevan, 1952, p. 75)

With the benefit of hindsight, it can now be seen that there were two serious problems inherent in Bevan's vision. The first was the impossibility of providing every member of the population with "the best that medical skill can provide," as a consequence of inevitable constraints on resources. This aspect has become increasingly important as a consequence of the subsequent development of expensive medical technologies, particularly in relation to drugs and surgical procedures (discussed later). The second was a consequence of the partial dependence of the health of a nation on the distribution of income within it, demonstrated recently by Wilkinson and Pickett (2009). Between 1979 and 1997, Conservative governments were in power and income inequalities increased sharply in the UK. In 1997, a "New Labour" government led by Tony Blair was more committed to market forces than to reducing inequality. Moreover, that government accelerated policies of privatisation to the extent that it became unrealistic to consider Britain as a "model" of a publicly provided national health service. Not only the principles of a policy, but the details of its implementation, have significant effects on its effectiveness. In Britain, there have been continuing and increasingly harmful deficiencies in the implementation of what it is contended here was originally a sound policy.

Barnett (1986) suggested that in 1940, members of the British cultural elite began work on what he characterised as "The Dream of New Jerusalem," to be built in Britain after the war was won:

> Selfish greed, the moral legacy of Victorian capitalism, would give way to Christian community, motivating men to work hard for the good of all. In this community, the citizen would be cushioned against the stab of poverty by full employment, welfare grants and pensions—all provided by beneficent state—from infancy to the end of earthly life. Universal free health care . . . would replace grim and run-down Victorian infirmaries and the ragged safety-netting of existing free medical services. (Barnett, 1986, p. 13)

But "the cost of realising this programme was to fall on a country with a ru-ined export trade, heavily in debt to its bankers (the Sterling Area Commonwealth countries and the United States)" (Barnett, 1986, p. 37). By 1945, "the British in their dreams and illusions . . . had already written the broad scenario for Britain's post-war descent to . . . the place of fourteenth in the whole non-communist world in terms of annual GNP per head" (Barnett, 1986, p. 304). This failure of gross national product (GNP) per capita, to grow as fast in Britain as in the United States and Germany may have been inevitable, however, as a consequence of Britain's deeply entrenched failure to adapt its education and vocational train-ing to the changing needs of the economy (Barnett, 1986, pp. 201–233; Senker, 1992, pp. 5–25). Indeed, I suggest that the election of a Labour Government to succeed the coalition led by Winston Churchill, the great war leader, represented in effect a decision to choose social welfare in preference to economic growth. Blair (1996, p. 10) disagreed, claiming that the 1945 Labour Government "laid enormous emphasis on economic modernisation as well as pursuing the goal of social justice."

I suggest that this history could have relevance for developing countries that lack some of the capabilities and resources necessary for pursuing rapid economic growth. Some such countries might well achieve substantial benefits in terms of welfare through adoption of collective policies in relation to health, even though several developed countries' health services have delivered better performance over time than Britain's in some respects, for example, those of France, Sweden, other Scandinavian countries and Japan. In Europe, the government is generally the insurer of health care for all citizens, and government insurance is financed by general taxation. Japan's achievement of the longest life expectancy in the world and its very low infant mortality rates can be attributed to some extent to its health care system (Tatara & Okamoto, 2009). The Japanese system is based on universal coverage of the population by statutory health insurance, but it con-centrates on disease treatment. This inhibits it from addressing health challenges associated with lifestyle-related diseases such as diabetes and hypertension. This contrasts with the British NHS in which, in principle, preventive and curative services should not be separated.

During the last sixty years, there has been considerable expansion in "the best that medical skill can provide," based largely on developments in science and technology (discussed later). In the United States, "the best that medical skill can provide" has indeed become available—but only to a small section of society—the wealthiest. But the countries that have performed best in terms of the general health of their citizens have been those who have provided the majority of their populations with more than a minimum of health care, either free or cheaply, financed by national taxation and/or insurance schemes, with national or nearly national coverage.

The effectiveness of national health care schemes in improving health has been the consequence not merely or even mainly of the effectiveness of the health services themselves, but also of numerous other factors, including the extent of equality and inequality in the distribution of income within countries or regions, diet and cultural factors. There is no clear evidence that private or public sector hospitals provide consistently better performance in terms of health or mortality outcomes (Milcent, 2005). Public ownership of hospitals and other health service provision is dominant in Nordic countries (Magnussen, Vrangbaek, & Saltman 2009). In both France and Britain, there is a mixed system of publicly- and privately-owned hospitals. In the British system, private hospitals may increase profits in effect by passing on costs to publicly-owned hospitals (Senker, 2011).

Wilkinson and Pickett (2009) provided evidence that, above a certain level, increases in income per head do not generally help secure improvements in health, such as increases in life expectancy and reductions in infant mortality (pp. 79–81). The failures of the British NHS to improve the health of the population to the extent that similar systems have been effective in other European countries were probably related partly to the extent of income inequality in Britain, rather than to specific features of the NHS compared with health systems in Sweden, Norway and France.

Saunders (2010) is critical of both Wilkinson and Pickett's statistical evidence and their causal arguments. For example, he suggests that the Wilkinson and Pickett claim that average life expectancy is linked to income inequality rests entirely on the longevity of people in Japan, which probably is related to diet, genes or a mixture of the two, and that if you take Japan out of the analysis, the apparent association of longevity with income inequality collapses. The (more equal) Scandinavian nations routinely appear at one end of many of Wilkinson and Pickett's graphs, and the (less equal) Anglo nations often appear at the other. According to Saunders, these differences probably reflect a deeper divergence between Nordic and Anglo cultures, for when we look beyond these clusters and search for evidence that might link inequality to social outcomes in other countries, we search in vain. Saunders maintains that *The Spirit Level* fails to show that health and well-being vary with the degree of income inequality in a country. But it does show that Japan and the Scandinavian nations tend to come out ahead of the 'Anglosphere' countries such as the UK, United States, Australia and New Zealand on a range of indicators. Saunders suggests that the most likely explanation lies in the history and cultures of these countries. Saunders' claims are all contested by Wilkinson and Pickett (2010).

This controversy may well continue. My tentative judgement is that so far, the challenges made by Saunders to the case made by Wilkinson and Pickett are unconvincing. In contrast to the United States, Norway, Sweden, France, Japan and the UK all have health systems that ensure that only small sections of the population suffer from severe deprivation. But most indicators of the health of the population, such as life expectancy and infant mortality, are rather better in Sweden, France, Norway and Japan than they are in the UK. Disparities between the health of sections of the population tend to be more extreme in countries with greater disparities in income—the poor health of poor people in those countries pulls down the average standard of health.

The United States' huge health expenditure per person is ineffective in improving the health of the whole nation for several reasons, which include extreme income inequality and also inequality in the distribution of health expenditure (Wilkinson & Pickett, 2010, pp. 79–81). The proportion of residents of the United States who are not covered by health insurance is large and growing. In 2000, 13.7 percent of the population was not covered by health insurance, and this proportion rose to 15.3 percent in 2007. Unemployment has risen recently, median household income is now falling and poverty is rising, so the proportion of the population not covered by health insurance is expected to rise to well over 16 percent, about a sixth of the population. The cost of health care is significantly increased by huge malpractice awards; in the attempt to avoid being sued, doctors practice defensive medicine, which is inevitably expensive and inefficient. Despite the highest per capita spending on health in the world, the health of the average US citizen is not particularly good. Average life expectancy is not particularly high, and maternal mortality figures are not particularly low. In relation to infant mortality, the United States has a worse record than much poorer countries, such as Singapore, the Czech Republic, South Korea and the UK (Blanchflower, 2009).

There were certain features of British government and society which facilitated success: The British government was relatively autonomous, despite the fact that NHS development was largely financed in effect by the United States through the Marshall Plan. The British government was competent to initiate and run a National Health Service and had, in effect, been instructed by its electorate to do so; the electorate was sufficiently confident in the government to entrust it with running a health service free to individuals at the point of consumption.

Other governments, particularly in developing countries and the United States, lack some such characteristics. Numerous developing countries, especially the poorest countries, lack governments with the authority and competence to implement such developments. The United States is peculiar in some respects. In effect, the people of the United States, by their election of President Obama, did in some sense 'authorise' such a development, although the recent successes of the Republican Party in winning Congressional elections indicate that many may be having second thoughts. While the US federal government has the competence to implement a national health service, distrust of government is so prevalent in culture and politics as to create strong barriers. Not only the strong resistance of vested interests such as drug corporations, insurance corporations and doctors, but also pervasive distrust of 'Big Government' amongst the electorate, forced President Obama to turn to insurance to facilitate implementation of the desire of a substantial proportion of the electorate for a more efficient and fairer system of delivering health services.

Health and Inequality in Developing Countries

The World Bank and the Wharton School have advocated voluntary health insurance as the best way of coping with developing countries' problems of health provision, as do the World Health Organization, Pan American Health Organization, and the U.S. National Institutes of Health (Waitzkin, Jasso-Aguilar, & Iriart, 2007). But recent evidence from the United States and Latin America briefly outlined here indicates that the majority of people in low- and middle-income countries do not have access to insurance coverage at the time of illness. Over 80 percent of the population in developing countries rely on out-of-pocket payments to providers when they need medical care. This exposes them to impoverishment when they suffer catastrophic illness.

Although some health system privatisation has occurred in Europe, more has taken place in Latin America than in other regions, and the most notable example of the application of these principles has been in the United States, where the country has produced an unfair and inefficient health service. Moreover, largely as a consequence of inequality, Latin American countries, while by no means amongst the poorest in the world, are still at a stage of economic development, where increases in prosperity might benefit the health of their population (Wilkinson & Pickett, 2009, Figure 1.1, p. 7). And their failure to develop their economies as rapidly as the leading European countries has itself been attributed to extreme income inequalities (Arocena & Senker, 2003).

Privatisation has not generally succeeded in improving access to health services for vulnerable groups in Latin America. Although its impact has differed among Latin American countries, expansion of private insurance has generally worsened access to needed services. Privatisation, either through conversion of public sector to private sector insurance, or by expansion of private insurance through enhanced participation by corporate entrepreneurs, has usually improved conditions for private corporations and has led to higher administrative costs. The adoption of privatisation and insurance schemes in Latin America has mainly been a consequence of policies of international organisations such as The World Bank and IMF, influenced by academics in the United States (e.g., in the Wharton School; Waitzkin, Jasso-Aguilar, & Iriart, 2007).

For developing countries, publicly run health systems are likely to offer better services to the community as a whole than systems based on private health insurance. But developing countries depend heavily on aid from external donors to provide the resources for improving health. There is an interesting contrast with Britain in the late 1940s. Britain was sufficiently autonomous to be able to divert aid provided by the United States—provided principally with the aim of helping the British economy to develop—to support the development and implementation of a National Health Service. Most developing countries lack sufficient control over the aid they receive to integrate it into effective comprehensive health services, and several lack strong enough governance to do this.

"The growth of private sector health care provision in developing countries has largely been a consequence of 'passive privatization'" (Global Health Watch, 2005, p. 65). The collapse of the public sector has led to the emergence of a disorganised, unregulated and even chaotic provider market in many developing countries, particularly at the primary level of health care.

The incapacity of public services has also resulted in governments and donors relying upon NGOs, UN agencies, charities, religious groups and humanitarian organisations to plug the gaps in public provision not only in primary care but also in essential hospital services and in response to humanitarian emergencies (Global Health Watch, 2005, p. 65).

Particularly in developing countries, issues relating to health are inextricably intertwined with issues relating to poverty. Collier (2008) suggested that the poorest billion people are living in "development traps." The conditions in countries locked in development traps have severe consequences for health. In particular, civil war involves mass rape and mass migration and this has probably been a significant factor in the spread of HIV/AIDS in Africa (pp. 5, 31).

Sachs (2005) has focussed on the consequences of health problems such as malaria and HIV/AIDS as trapping the populations of some countries in poverty, although there is not yet an adequate explanation of the extraordinary high rates of transmission of HIV/AIDS in African countries. He produced proposals for dealing with the poor health and poverty endemic in developing countries. His plan to end poverty in our lifetime is based on 'enlightened globalisation' to be created by a mass public movement toward the creation of a global economic system, which would address the needs of the poorest of the poor and the global environment (Sachs, 2005).

The principal instrument in Sachs's plan would be development aid by rich countries to be increased enormously to 0.7 percent of the GDP over twenty years (Sachs, 2005, p. 345). He looks to the United States to lead this process of creating enormous aid programmes that can best be stimulated by 'coalition building' led by the president. In relation specifically to health, Sachs considered the five-year, $15 billion President's Emergency Plan For AIDS Relief (PEPFAR), in which President Bush asked Congress for $15 billion a year, including $10 billion in new money to be spent on controlling AIDS in the most afflicted nations of Africa and the Caribbean. This comprehensive plan was designed to prevent 7 million new AIDS infections, to treat at least 2 million people with life-extending drugs (mainly HAART) and to provide care for sufferers and orphaned children. The coalition that led to this plan being adopted by President Bush included experts from the National Institutes of Health, celebrities, religious leaders, liberals and conservatives (Sachs, 2005, pp. 343–345.)

Experience indicates that caution needs to be exercised in considering prospects for success of such plans. For example, Davis and Flowers (2009) pointed out that access to HAART (Highly Active Anti-Retroviral Treatment [of HIV/AIDS]) is restricted by its cost and the difficulties of providing it in different health care settings throughout the world. In 2003, UNAIDS initiated a global campaign to increase the number of people receiving treatment to 3 million by 2005. By June 2005, it was estimated that 1 million people were using HAART in resource-poor settings, up from 440,000 in 2003, but far short of the campaign target. In transitional countries, treatment-related uncertainty was related to emerging but fragile economic systems, where expensive HIV treatments can be funded only periodically. Until recently in Serbia, patients could not expect to have constant access to HAART treatments. For some months in a year, there were restrictions in funding for health care (Davis & Flowers, 2009).

The relationship between malaria and poverty runs both ways—a kind of vicious circle. Richer households and governments can afford to spray homes with insecticide, can afford to install screen doors to help keep mosquitoes from entering, can afford insecticide-treated bed nets and can ensure access to health care and effective medications when needed. Repeated bouts of malaria result in poor performance and school attendance; they also cause parents to have more children to replace those that die (Sachs, 2005, pp. 197–198).

The malaria problem is more grave in Africa than in India, partly because mosquitoes in Africa tend to bite humans rather than cattle, while in India they are more likely to bite cattle. But the amount of rich country help to Africa to fight malaria is minimal in relation to the size of the problem. "The IMF and the World Bank were apparently too busy arguing for budget cuts and privatisation of sugar mills to have much time left to deal with malaria"(Sachs, 2005, p. 200). Easterly (2005) acknowledges the success of many international aid efforts, from the dissemination of deworming drugs and the use of oral rehydration therapy for diarrheal diseases to indoor spraying to control malaria and several programs to slow down the spread of AIDS.

In relation to basic health, Sachs's proposals envisage village clinics, each with one doctor and one nurse who might provide a range of essential health services, including skilled birth attendants and sexual and reproductive health services, together with services specific to each area. For example, in regions where malaria is endemic, free antimalarial bed nets and effective malarial medicines

Peter Senker

would be provided by village clinics. In regions in which HIV/AIDS is endemic, village clinics would provide treatments for opportunistic infections associated with HIV/AIDS together with anti-retroviral therapy for late-stage AIDS (Sachs, 2005, p. 233).

Innovation and Inequality

The basic medical research that underpins drug development may be funded by government, industry or medical charities. The drug development process-testing, manufacturing and marketing are primarily funded by the pharmaceutical industry that profits from the fruits of the whole process, from basic medical research to marketing. Drug development builds on basic science and research and development. In order to recoup these costs, pharmaceutical companies seek to gain market exclusivity by patenting new compounds. On average, the effective patent life of a new drug is around twelve years. This is a case of public funds initiating the creation of intellectual property rights. Sharpe (2009, p.50) suggested that, "Public investment in a capitalist society is often made for private gain," as in the case of pharmaceuticals.

Innovation has been rapid in terms of development of drugs and other medical treatments but not as rapid as had been assumed until recently. "Rather than producing revolutionary changes, medicinal biotechnology is following a well-established pattern of slow and incremental technology diffusion" (Nightingale & Martin, 2009, p. 159). More important, technological innovation is not taking place in directions that would enable it to make the contribution to the health of the world's population of which it could be capable.

A survey of new drugs registered between 1975 and 1999 with US and EU regulatory bodies shows that the majority target diseases that are prevalent in developed countries and those that drug companies' current research programmes concentrate on, such as diseases of the central nervous system (15.1 percent), cardiovascular disease (12.8 percent), cancer (8 percent), and noninfectious respiratory conditions (6.4 percent). Only a small proportion of new drugs targets diseases that are largely confined to developing countries. For instance, 0.3 percent of all new drugs were targeted at malaria; of 1,393 chemical entities taken to market between 1975 and 1999, only 16 were for neglected diseases, malaria and tuberculosis. Some important initiatives were taken in the first five years of the new millennium to tackle previously neglected diseases, which afflict developing countries, such as malaria, HIV/AIDS, sleeping sickness, leishmaniasis and Chagas disease (Parliamentary Office of Science and Technology, 2005, p. 3).

Billions of dollars are spent annually on medical research and development, but only a small fraction of these dollars are devoted to diseases such as tuberculosis and malaria: only 20 of the 1,500 new drugs approved by the US Food and

Drug Administration over the past twenty five years were for the diseases that disproportionately affect people in developing countries. According to the World Health Organisation (WHO), 30 percent of the world's population lacks access to life-saving medicines. In some countries in Asia and Africa, the number may be as high as 50 percent. The international community has been encouraging generic competition in the attempt to drive prices down and the adoption of tiered pricing to create situations in which higher prices in developed countries subsidised drugs for the world's poor. But some generic manufacturers have made and distributed substandard drugs that have threatened public health and encouraged drug-resistant strains of pathogens. Some pharmaceutical companies have only adopted tiered prices for high-profile HIV drugs. Even when manufacturers have lowered their prices, patients pay far too much as a consequence of taxes, markups by middlemen and corruption (Bate, 2008).

Policies in developed countries, notably the United States, fail to contribute sufficiently to the promotion of technological innovation in relation to the diseases that affect developing countries disproportionately: US policy is "more concerned with achieving greater market access and competitive edge in developing country markets, and does not take adequate account of the priorities of developing countries with respect to public health" (Musungu & Oh, 2006). Many techniques for diagnosis and treatment currently in use in developing countries are cumbersome and unsuitable for use in areas that lack running water, refrigeration or electricity. Vaccines are critical to disease management in such areas. The enormous expense of refrigeration needed to maintain the required temperature can add up to 80 percent of the cost of vaccine delivery in developing countries. Most vaccines and many drugs are administered by injection, and many new cases of blood-borne diseases, such as HIV/AIDS and hepatitis B, are caused each year by unsanitary injections.

Numerous technologies are available that could offer better, more effective treatment for diseases prevalent in developing countries. But insufficient resources are devoted to development and exploitation of such technologies. Alternatives to injections, frequent dosing and refrigeration could increase safe access to drugs and vaccines, saving millions of lives. There are needs for powdered vaccines, edible vaccines and controlled-release formulations that replace the need for multiple doses. Several molecular diagnostic technologies are either already in use or are being tested in such regions. Efforts to produce new vaccines, for example against malaria or hepatitis B, often involve recombinant technologies. Some recombinant vaccines are already being manufactured in developing countries at a fraction of the cost of imported alternatives. It has also been suggested that female-controlled protection against sexually transmitted disease, in particular HIV/AIDS, might be assisted by genomics-based technologies (Daar et al., 2002).

Local small companies, particularly in China, India and Brazil, have made significant contributions to local and global health needs through low-cost manufacturing of health products. Until recently, several medical devices—especially expensive, complex machines such as CT scanners—were only available for the benefit of patients in the developed world, together with the very richest patients in developing countries. However, application of the principles of 'frugal innovation,' a concept pioneered by Prahalad (2006) in India, has led to the development of much lower cost products capable of carrying out similar functions. For example, Zhongxing Medical, a small medical devices company in China, has developed an x-ray machine that costs only about 5 percent of the cost of a typical x-ray machine made in developed countries. So as to achieve low cost, Zhongxing designed its machine so that it could only perform routine chest scans, but these represent the vast majority of scans patients need (Sehgal, Dehoff, & Panneer, 2010). Several subsidiaries of multinational companies have also become involved. For example, Philips has used frugal innovation principles to develop a bedside patient monitoring system in China (Zeschky, Eidenmayer, & Gassman, 2011, p. 41). Indeed, Philips has acquired several Chinese companies that have developed medical devices using frugal innovation, partly in order to participate in the Chinese drive to extend health care into rural areas. Indian and Chinese use of frugal innovation to develop cheap but effective medical devices is expanding very rapidly (*The Economist*, 2011).

Similarly, numerous small biotechnology firms have been created in India, China, Brazil and South Africa to provide affordable drugs and vaccines for tackling diseases such as tuberculosis and tropical diseases prevalent in developing countries.

Major patent holders such as GlaxoSmithKline are increasingly willing to share intellectual property rights in relation to treatment of the diseases of poverty. GlaxoSmithKline will also publish its research results related to a group of over 13,000 promising compounds against malaria.

Combining the innovative capacities of firms in the emerging economies with increased access to knowledge and technologies possessed by multinational corporations could accelerate development of products designed to treat the diseases of the poor. But drugs and vaccines are expensive to test and bring to market, so small innovative companies in developing countries often go into partnership with large pharmaceutical companies. It seems possible that market forces will push such firms and partnerships toward concentrating on development and production of drugs and devices to tackle the diseases of richer countries. For example, a small company in India is now working with a Danish firm on a treatment for diabetes, and another in China is working with US firms on drugs for inflammatory bowel disease. An Indian company has developed hepatitis B vaccines, given to millions of Indians, by using bacteria equipped with viral genes, reducing vaccination costs enormously. But it was bought by a French company in 2009, and market forces could well shift its focus away from treating developing country diseases (Rezaie & Singer, 2010).

New Horizons

Where all aspects of health care have been concentrated in private hands, most notably in the United States, this has resulted in the development of powerful ranges of pharmaceuticals, scientific tests and equipment and surgical procedures, but also in a very inefficient and unfair health care system. US expenditure on health care per capita is immense, but this has resulted in a population whose average levels of health are poor relative to several other countries in which health expenditure is far less but where the distribution of both income and health care is much less unequal.

Improving the health of the poorest billion people living in countries caught in development traps mainly in Africa may be largely a matter of improving governance by ending civil wars and of creating economic growth in those countries (Collier, 2008). These issues are most important but lie outside the scope of this chapter. Until such measures are successful, the prospects of establishing effective health care systems in such countries are inevitably poor.

For richer countries, whether they are still developing countries or whether their average income has risen sufficiently for them to be classified as developed countries, fundamental problems are those of relationships among, and assigning priorities among, economic growth, equality, inequality and the development of fair and efficient health care systems. These priorities are best established on a democratic basis by the people themselves on as local a basis as is feasible, not by international organisations and aid donors.

The United States has had a strong influence over the policies of international organisations such as the International Monetary Fund and The World Bank in relation to developing countries. This influence has ensured undue dominance of priorities of economic growth over reduction of inequality and over social expenditure on health and education. Moreover, it has been suggested that often IMF and World Bank Policies have been misguided even in terms of promoting economic growth (Stiglitz, 2002). Foundations in the United States have also tried to influence developing countries, especially in Latin America, in favour of financing health service through voluntary private insurance. Voluntary private insurance has not proved to be satisfactory as a basis for financing health service in the United States, and it has proved to be even less satisfactory in Latin American countries.

While there has been some progress in inequality reduction, perhaps especially through the use of frugal innovation applied to low-cost medical devices and machines, the situation more than ten years ago remains fundamentally un-changed. Even in cases where innovation is initially directed at solving the health problems of the poor in developing countries, market incentives often result in the subsequent diversion of innovations to treatment of richer people, accompa-nied by much larger resources.

SEVEN

Elearning or E(l)earning

Contemporary Developments in the Commodification and Consumption of Education

Allyson Malatesta

"I see a lot of crisis management, but I don't see any horizon creation." (Burg, cited in Rees, 2001, p. 33)

The statement from Avraham Burg, although made in reference to the *intifadeh* in Israel, resonates with current developments in teaching and learning with new technologies, as the use of these technologies in education is hyped as the answer to many of the world's problems. If there is a problem to be solved and education is posited as the means by which this can be achieved, then the necessity of using technology to facilitate this will be advocated. However, the solutions often turn out to be quick fixes or 'crisis management,' with no long-term benefits or strategies in place.

Technologies have always been important in education, and one of the earliest technologies to make a significant contribution was the Gutenberg Press. A combination of invention and innovation, this early technology facilitated the mass production of affordable books that could be acquired by literate individuals (Postman, 1986, cited in Evans, 1998). If you were not literate then, you were effectively excluded from the educational benefits that these books might afford you, thereby giving rise to a state of educational inequality. In the 1800s, some four hundred years after the appearance of this revolutionary technology, a geography teacher in Scotland, tired of the necessity of writing problems on pupils' individual slates, gathered them up and mounted them on the wall at the front of the classroom (Woods, 2007). Thus was born the blackboard, and this, thereafter, enabled whole class teaching to take place. Subsequently, the blackboard developed into the whiteboard and was further innovated to become the interactive whiteboard. The new tablet computers, such as the Apple iPad, when switched off, closely resemble those slates from years ago. If a number of them were fixed to the wall, they would make a reasonable semblance of a blackboard. Switch them on and you have an interactive whiteboard. Technology would appear to have come full circle for those able to engage with twenty first-century technologies. However, in many developing countries, the original slates and blackboards are still very much in use in educational settings, and as can be seen from the following discussion, it is the most appropriate technology for each individual situation and purpose that should be the technology of choice, if we are to address issues of sustainable development through education.

There have been many and diverse developments in educational technology and methods of teaching and learning, of which those involved in the provision of education have never failed to take advantage, often promoted by a rhetoric of reducing inequality. However, as can be seen throughout this chapter, the gap between the educationally privileged and underprivileged shows no signs of narrowing and actually may be widening ever further. Developments in teaching and learning have been put forward as means to bring education to all parts of the world as part of the rationale inherent in creating a 'global village' and bridging the 'digital divide.' However, in examining some of the current changes in educational provision, increasing privatisation and marketisation and the role of technology in this process, this chapter argues that the promotion of technology as a way of dealing with the problems of equity and diversity globally has been driven by a misplaced belief that technology itself can be the change agent coupled with an increasingly capitalist agenda, and thus these approaches too often fail to deliver the much-hyped promises.

Privatisation and Marketisation of Education

Before considering the way in which teaching with technology has changed the way education is delivered and accessed and how this has further facilitated the commodification of education, it is necessary to consider the way in which the provision of education itself has changed. Ball, in his discussion of the challenges for education and social policy-making, considers two possible responses that policy-makers might lean toward and which were put forward by Brown and Lauder (1996), that of neo-Fordism and post-Fordism. According to Brown and Lauder, neo-Fordism "can be characterized in terms of creating greater market flexibility through a reduction in social overheads and the power of trade unions, the privatisation of public utilities and the welfare state, as well as the celebration of competitive individualism" (Brown & Lauder, 1996, cited in Ball, 1998, p. 121). Post-Fordism can "be defined in terms of the development of the state as a 'strategic trader' shaping the direction of the national economy through investment in key economic sectors and in the development of human capital" (Brown & Lauder, 1996, cited in Ball, 1998, p. 121). Commenting on this, Ball (1998) stated the following: "This policy dualism is well represented in contemporary education policies which tie together individual, consumer choice in education markets with rhetorics and policies aimed at furthering national economic interests" (p. 122).

A decade later and taking the UK as an example, Ball (2008) discussed the policies of New Labour and Tony Blair's school reforms, the aims of which were to "make this country at ease with globalisation" (Blair, 2005). This process of marketisation was begun under the Thatcher governments with the 1988 Education Act, which introduced the market through the vehicle of increased 'parental choice.' Blair maintained that privatisation was necessary in order to transform education and Labour's Education White Paper proposed opening up the system to new providers and new partners with the aim of providing greater choice. It is in this direction that education has steadily been moving, with the private sector becoming increasingly involved in the provision of education (and other public services [Whitty, 2000]) with the promises of greater choice, but at what cost? As Ball (2004) stated, although there are benefits to be obtained from some forms of privatisation, the benefits are sometimes exaggerated, while the costs (primarily social costs) are neglected (p. 2). He also went on to point out that the role of the profit motive is left out of the equation and business failures and business ethics are also neglected. It can be seen later the extent to which there have been failures in businesses involved in the provision and/or delivery of education, in this case online education, as that is the focus of this chapter. It is the availability of the rapidly evolving electronic technologies and the increasing reach of the Internet that are enabling these changes in educational provision to take place, but there are clearly consequences of their use for all involved.

The discussion of the marketisation of education has been primarily concerned with face-to-face teaching and learning; however, utilising new media technologies has enabled learning to be further commodified, and courses can now be designed that draw on content from different providers and delivery from a diverse range of institutions. Marx put forward two classes of commodities, that of labour power and that of commodities as distinct from labour power itself (Marx, 1863, cited in Rikowski, 2007). Hatcher and Hirtt (1999, cited in Rikowski, 2007) saw the former as the business agenda for education (the production of labour power as human capital for businesses) and the latter as the business agenda in education (the process of the business takeover of education for profits). There have been criticisms of the commodification of education whereby education, as part of cultural capitalism, is treated as a private gain as opposed to a public good, a tradable commodity, and learners are referred to as customers, clients or consumers. As Willmott (1995, cited in Ball, 2004) stated, "Students have been explicitly constituted as 'customers,' a development that further reinforces the idea that a degree is a commodity that (hopefully) can be exchanged for a job rather than as a liberal education that prepares students for life" (p. 5). Seeing education as a tradable commodity has, however, contributed to falling educational standards, according to Noble (1997). Indeed, using the UK as an example, the National Grid for Learning (DfEE, 1997) was hyped as the answer to preparing children for the world of tomorrow utilizing the potential of new technologies (Wellington, 2005). Yet the current debates are still focusing on the numbers of children leaving school with low levels of literacy and numeracy, despite the widespread integration of ICTs at all levels of schooling and into all subjects.

Whether education can actually be seen as a tradable commodity is disputed by various bodies. The European Students' Union (ESU; previously The National Unions of Students in Europe [ESIB]) emphasised the important role of education as a means for social development, democratic empowerment, the general well-being and economic competitiveness of societies, knowledge accumulation and sharing, cultural capital and as a means for personal growth. The union maintained that because of these roles, education could not be reduced to a mere economic, tradable commodity (ESIB, 2002, p. 2) and declared that "The notion that education is a commodity tradable with the same rules as any commercial product is unacceptable" (ESIB, 2001). The union returned to this debate in 2005 and again stressed ". . . that open access to all levels of education is the cornerstone of a socially, culturally and democratically inclusive society . . . ." The union also commented that in the economic debate, which emphasises the importance of a knowledge-based economy, this definition is contested and education is understood solely as an economic factor rather than a tool for social development (ESIB, 2005). The 1998 UNESCO Declaration stated that, "Higher education exists to serve the public interest and is not a commodity" (Fouilhoux, 2004), and again in 2008, UNESCO reinforced this by stating that, "Education is a public good and a human right from which nobody can be excluded since it contributes to the development of people and society" (UNESCO, 2008, p. 6). However, the World Trade Organisation (WTO) believes that globalization and education are related and they facilitate globalization by opening up public services (including education) to international capital (Rikowski, 2002a). Monbiot (2001, cited in Rikowski, 2002b) and Lucas (2001, cited in Rikowski, 2002b) pointed out that the practical results of further opening the markets of poor nations to transnational corporations are likely to be greater inequalities between rich and poor nations. It is clear, however, despite these views, that education in the form of elearning is being widely marketed and 'consumed' in many areas of the world, and close scrutiny of the quality of this education needs to be prioritised for a number of reasons, not the least of which is whether the learning outcomes that are promised are being fulfilled and whether this leads to the creation of a better society.

The use of new technologies in education is also often promoted for its capacity to enable learners to choose from a variety of content and providers and construct their own path of learning. It also means that learners will be able to choose from offerings and providers globally but also from organisations that may be unscrupulous or inexperienced and provide online education or training of dubious quality. In this regard, some governments are seeking to protect their students and home providers by excluding overseas providers from offering education and training services. This situation is being redressed by the WTO in the General Agreement on Trade in Services (GATS), which aims to reduce barriers and enable education and educational providers to operate freely among countries. Scherrer (2005) sounded a warning note with regard to GATS' long-term strategy for the commodification of education. He pointed out that GATS is not an instrument for pushing through neoliberal reforms but ". . . can be used to secure the power of capital in the long term by privileging private owners of educational services in relation to the public and to the actual providers of these services, the faculty" (p. 485). This has repercussions for learners who, whilst having greater choice, will have to determine the quality and standards of the education being provided. Being able to make these choices wisely will be of increasing importance to learners in the current climate of lifelong learning and continuous professional development. It is not always apparent who, exactly, is offering some of the online courses that are available. Unless a learner carries out some detailed research, which few people are likely to undertake before embarking on a chosen course, there is no way of knowing how reliable and responsible an online provider might be. It is easy, on the internet, to 'hide' or misrepresent information relating to the accreditation of providers or to mislead people as to the type of organisation offering courses or even to suggest links to well-known and reputable higher education or other institutions, which might be tenuous at best or nonexistent at worst. It is the availability of the new technologies and the marketisation of education that have enabled private, for-profit organisations to jump on the bandwagon of elearning to the extent they have at the present time.

Technology as the Machine of Educational Change

In the early days of educational technology, when computers had only been in use for teaching and learning for a relatively short time, Heaford (1983), with some forethought, stated that, "Computer Literacy—at any level, at any age, at any price—is the fundamental basis upon which the future of industry, and maybe society, will depend" (p. v). He tempered this by also saying that there was (at the time) ". . . a predominant concern with finding a market for the technology, or finding a problem for the solution" (p. v). To date, nothing has changed; the same claims regarding computer literacy are still being made, and technology suppliers are constantly looking for new markets for their products solely to increase their profits. The only concrete changes concern the increase in computing power and the speed of innovation and development of the range and diversity of ICTs.

Many involved in the education sector have consistently looked to the development of new technologies, especially ICTs, to see how they might be utilised in or impact on teaching and learning. This is especially pertinent in the realm of distance education. From the early days of The Open University's printed materials through other media such as TV and radio broadcasts, video, audio, CD-ROMs and DVDs, it is only since the advent of the internet that new ways of distance learning, using computer-mediated communication (CMC), have been possible. Variously known as computer-mediated learning, online learning or el-earning, it utilises the facilities that are offered by the new media technologies. John Chambers (2000, cited in Alexander, 2001), CEO of Cisco systems, made a telling statement in suggesting that, "The next big killer application for the Internet is going to be education. Education over the Internet is going to be so big it is going to make email look like a rounding error" (p. 2). It was clear that technology providers would be at the forefront of this drive toward online education, for the economic benefits they would derive from it.

Current educational thinking maintains that, as new ICT technologies enable learners to collaborate with others online, this would lead to a better quality of learning and also bring together learners from all parts of the globe. The rhetoric surrounding the use of new technologies for teaching and learning also promised to bring greater equity globally between the information 'haves' and 'have-nots' and also to bridge the digital divide. Referring to developing countries, where these information and technologically disenfranchised citizens reside, this promise has been couched in terms that view the world's four billion poor as a market, not a burden, and, therefore, as potential customers, not just human beings with needs (Prahalad, 2004, cited in de Miranda, 2009, p. 24). However, Norris suggested that the global pattern of inequality in internet use mirrored the pattern of access to earlier ICTs and that the disparity of internet access between developed and developing countries is not particular to the nature of internet technology but due to deep-rooted and endemic contextual factors within those societies (Norris, 2001, cited in Avgerou & Madon, 2005, p. 6). Nevertheless, as education is the means by which the poor and disadvantaged are able to improve their situation, governments are constantly seeking ways in which they can provide this, as improvements in people's lives at a micro-level will ultimately benefit the country at a macro-level. ICTs have been promoted as the way in which these aims can be met and education made available to diverse and widespread communities. Producers of ICTs are never slow to take advantage of new markets for their products, and tying the promotion of these to the provision of education ensures their motives are seen as altruistic rather than purely economic.

The increasing use of media technologies, which are seen as an even greater enabler in the marketisation of education, led Levidow (2002) to comment that, "In the ruling ideology, marketisation imperatives are attributed to inherent socio-economic qualities of ICT" (p. 227). Some question that ICT alone can create the conditions of equality and the satisfaction of needs for all human beings. However, an interesting point was made by Wyatt (2008), when she "challenged the digital imperative" by commenting that some people actively choose not to participate in the so-called "information society" (p. 9). Arora's (2010) detailed research undertaken in the Central Himalayas clearly demonstrates that not all people feel the need to participate in the digital world and so-called information society. The insistence on policy-makers and technology suppliers advocating so strongly the necessity of ICT in all spheres of daily life is an example of technology push, which comes in the main from manufacturers of technology equipment. Once technology companies have pushed their products onto the market, they will then have a ready and passive market for costly and seemingly necessary, if not essential from their perspective, upgrades. However, despite this constant push from technology manufacturers, it is clear that some users are making up their own minds as to whether they have a use for ICTs at all or which technologies in particular or to what extent they can utilize these technologies. It is imperative, if issues of equality and diversity are to be addressed, that these choices are made wisely by those concerned in the direct provision of education.

Equality, Diversity and Elearning

Learning with new media has been seen as an important means for addressing issues of equality and diversity in learning. This can encompass providing alternative ways of learning and content in different formats to address the needs of learners with regard to gender, ethnicity and a variety of disabilities. Elearning is also promoted as providing the opportunity for individualization of consumption and 'personalisation' of content and varied methods of content delivery; this is not as easy and straightforward to achieve as it is posited. There is now a significant body of literature that focuses on the concept of learning objects (LOs) and repositories, that is, databases, that can (supposedly) enable courses of learning to be constructed using individual LOs. An LO is defined by AOL as "a portion of a course packaged with sufficient information to be reusable, accessible, interoperable, and durable" (SCORM, 2001, cited in Weller, 2004, p. 293). However, there has been much debate about the efficacy of this approach and whether a course designed in this way could ever be personalised to suit an individual learner. Even if the content of courses was able to be personalised, if it was part of an assessed course, providers would have to set rigid parameters regarding coursework, assessment deadlines and marking criteria. McGovern (2003) gave three reasons why personalisation would not work: Users did not have a compelling reason to personalise, the cost of doing it well outweighed the benefits for the producers and it had been seen as some Holy Grail of content management. Although his comments were made in relation to the personalisation of websites, they are equally valid when applied to education. If a learner were to personalise his or her course, how would that learner know which LOs to select in order to ensure that he or she learned what was necessary? What kind of qualifications could be offered if random choices as to the content of courses were made by the learners themselves? Online courses, if well designed, have been documented as being much more expensive to create and deliver, and providers are often more concerned with uploading the content for courses than considering the pedagogy underpinning it. Personalisation also serves to widen yet further the gap between equality and inequality, those who have choices as to how they can educate themselves and those who do not.

When the availability of online courses from private companies became more widespread, there was, it has to be said, considerable interest from learners. Many people, seduced by the promises of the flexibility offered by elearning, enrolled in courses covering a myriad of subjects. However, for many of these learners, it has been a less than edifying experience. To be a successful elearner requires a high degree of self-motivation, discipline and the ability to cope with feelings of isolation, uncertainty and confusion (Hara & Kling, 2001, p. 68). Learners can also experience feelings of frustration with the learning materials, methods and technology. Overall, it is a vastly different experience to being taught face-to-face. Most courses are delivered in English, which is an added complication for learners whose first language is not English, and if courses include a requirement for a significant proportion of online discussion, it often leads to misunderstandings among participants. In addition, often there is no way of knowing what a learner will receive in the way of course materials or support, and there are marked differences in what is provided by, for example, an institution such as The Open University, which has years of experience in distance learning, and a new or recent entrant to the market, which perhaps is only concerned with its revenue stream. This situation will not be resolved until there is regulation of private training, which will ensure that those who offer elearning courses are held accountable as reputable, reliable, education providers.

Despite the many promises of elearning for promoting equality through greater choice for consumers, however, the costs associated with elearning puts it out of reach of those who cannot afford to pay. Courses delivered wholly or partly online are more expensive than those delivered face-to-face. This is due to the high cost of development of good quality content, the costs associated with maintaining any virtual learning environment that might be used, and the additional costs of providing content in alternative formats, which any good course should aim to deliver, if it is to address the issues of equality and diversity. Should the course include discussion fora, which need to be moderated by tutors, this will also add to the overall costs. Moderating is extremely time-consuming for teaching staff and not as cost effective as teaching face-to-face. There are also many other costs that need to be taken into account, for example, those allied to the piloting of the course and training staff. Weller (2004, p. 294) stated that, "The cost benefits of creating e-learning courses are not as great as once envisaged," and cited Fielden (2002, cited in Weller, 2004, p. 295), who claimed that online instruction costs more than traditional instruction and that if development time is fully costed, the extra cost will be substantial. The high cost of developing online courses is another reason why LOs were seen to be a valuable way of bringing these costs down, as they are reusable and can be utilised in many different courses, rather than the content just being restricted to one single course.

In addition, complaints about the quality of courses and course materials have risen significantly in the UK, with Trading Standards receiving over 4,000 complaints, a 61 percent increase over the last twelve months (BBC News, 2010a). Thousands of learners complained that the courses cost too much, with some being tied into contracts, which included a credit agreement, pushing the overall cost of the course up considerably. In some cases, providers have gone into liquidation before courses have been completed, and learners are left still owing money but with no hope of being able to complete the course and receiving any qualification toward which they might have been working. Online learning has not proven to be as popular as some companies envisaged, and in a relatively short time, some of these providers have found this out, to their detriment. It is clear that the rhetoric concerning the reduced costs of elearning when compared to face-to-face learning is misplaced, as there are many extra costs that need to be taken into account, not just in the overall design of the courses but in maintaining the delivery platforms and associated technologies. Nevertheless, this does not present any barriers to unscrupulous providers who see elearning as a money-making opportunity, although they would do well to take note of past failures.

Many organisations, both in the state and private sectors, have ventured into the elearning marketplace with unfortunate and sometimes financially devastating consequences. In the UK, eUniversities Worldwide was a government-backed, public–private initiative for providing online courses to both home and overseas students in order to widen access to higher education. It was set up with £62 million of government funding in 2001 but closed in 2004 (UKeU, 2002). The former chairman of the UKeU holding company, Sir Brian Fender, following the closure, stated that, " . . . there was probably too much hype. There was some feeling that digital technologies would move faster than they have" (cited in Samuels, 2005). A task force set up to consider the elearning situation found that although there had been considerable investment in technology, the use of elearning was patchy, and that despite millions of pounds being spent, the status of UK public sector online learning was potentially worse (Samuels, 2005).

The situation in the US is no different. Hafner (2002, p. 2) stated that according to Eduventures, American universities have spent around $100 million on courses delivered over the World Wide Web, and several universities have now withdrawn from the elearning marketplace. This would seem to indicate that the perceived interest and demand for elearning at the tertiary level has not materialised either in the UK or US. However, as is always the case, new opportunities are constantly being sought, and education providers are not known for dwelling on past misfortunes but look for new ways to attract and engage learners. The widespread use of mobile technologies, especially mobile phones, offer them a new delivery medium as a conduit for their educational products. However, the usefulness of this technology for learning is seen very differently in developed and less developed countries.

From Elearning to Mlearning:
Bridging the Digital Divide in Developing Countries

Before discussing the move from elearning to mlearning (learning via a mobile phone), it would be useful to consider the situation for newly developing countries amid the push from technology providers to get these countries to adopt their products. The fact that elearning has been posited as a way to bridge the 'digital divide' and promote greater equality on a global scale fails to acknowledge the fact that many countries, especially developing countries, are not in a position to engage with digitised learning for a variety of reasons. Countless initiatives, arising out of international conferences and forums, have been put forward to address ways in which this divide can be bridged. However, despite the rhetoric, it is never clear in what way these initiatives will be implemented in practice. In December 2002, the United Nations General Assembly declared 2005 to 2014 as the UN Decade of Education for Sustainable Development (DESD), and Director-General Koïchiro Matsuura (cited in UNESCO, 2005) stressed the following:

> Education—in all its forms and at all levels—is not only an end in itself but is also one of the most powerful instruments we have for bringing about the changes required to achieve sustainable development. (p. 3)

As part of this initiative, UNESCO published a document entitled, 'Quality Education, Equity and Sustainable Development: A Holistic Vision Through UNESCO's Four World Education Conferences 2008–2009,' (UNESCO, 2008) which explained the key challenges for today's world, where education could make a difference, and which the four conferences would be tackling. It also stated that the aforementioned was dependent on policies being developed that were guided by a holistic vision of education systems. The document included the following extract from the UNESCO Medium Term Strategy 2008 to 2013:

> Development and economic prosperity depend on the ability of countries to educate all members of their societies and offer them lifelong learning. An innovative society prepares its people not only to embrace and adapt to change but also to manage and influence it. Education enriches cultures, creates mutual understanding and underpins peaceful societies. UNESCO is guided by upholding education as a human right and as an essential element for the full development of human potential. (UNESCO, 2008, p. 3)

The shared vision of the four conferences was one of education systems that encourage equity and inclusion, quality learning, flexibility and innovation. The document also stated that if education systems are characterized by inequality, discrimination and exclusion, they contribute to an increase in the existing social and economic disparities and also deviate from the path of equitable and sustainable development. It is clear from this that education needs to be delivered in a way that is the most appropriate for the circumstances prevailing. Interestingly, in a section in the DESD document headed, 'Educate Through Information and Communication Technologies (ICTs),' it states that local radio and internet access are both a means of training and exchange for solving community problems; the interesting part is the fact that the word 'radio,' a significantly older technology, appears before 'Internet.' The fact that the internet is not seen by everyone as the most appropriate medium for the delivery of education in developing countries is confirmed by a review, in Sub-Saharan Africa, of 150 distance learning education programmes which drew the conclusion that traditional, paper-based means of distance learning is more widely used, reliable and sustainable than online and web-based methods (Leary & Berge, 2006, cited in Gulati, 2008, p. 5). Some explanation for this can be gleaned from the results of a survey by Unwin (2008) of elearning in Africa and also research undertaken by Hollow and ICWE of 147 elearning practitioners from 34 African countries. Hollow's research provided some useful insights into the problems experienced or anticipated by the respondents to his survey. These included teachers not being sufficiently trained in the use of elearning to be able to use it successfully, the lack of funds available to implement planned programmes and issues allied to the high cost of bandwidth. Also mentioned were the initial startup and maintenance costs. Hollow (2009) himself concluded the following:

> The overall rationale for eLearning in Africa is still overly grounded in technology-driven agendas. There are encouraging signs that pedagogy is being increasingly prioritised but sustained work is required to ensure that the potential of eLearning continues to progress beyond simply training for ICT and focuses instead on educational outcomes. (pp. 9–10)

The findings of Hollow's survey are also reiterated in a short response to a news article by Mathy Vanbuel (2008) on why Africa cannot afford to miss the knowledge revolution. The response came from Benson Nindi, who stated that one of the major challenges for ICT in education could be summed up in three words: connectivity, access and affordability. He also stated that some people were being left behind and that a lack of coordination and the negative influence of self-ish business interests were complicating the challenges (Nindi, 2008, in Vanbuel, 2008). The use of ICT for learning has also been a challenge for Egypt. Kamel (2010, p. 187), referring to the country's literacy rates, stated that a substantial proportion of the country must overcome barriers beyond the merely technologi-cal in order to take advantage of the internet. The Smart Schools Network (SSN) was established to diffuse computing literacy by making computers available to students from an early age, and the initial, successful phase has been expanded with financial support from the United States Agency for International Development (USAID). The Egyptian Education Initiative (EEI) is a public–private partnership launched in 2006 between the government, the World Economic Forum (WEF), the IT Community and various ICT multinationals and organisations. The bi-lateral agreements that have been signed are with Microsoft, Intel, IBM, Oracle, Cisco, Computer Associates, HP and Siemens in addition to local partners. Kamel commented that, "The degree to which introduction of advanced ICT actually impacts teaching methods and learning outcomes also remains to be evaluated" (p. 190), but that the SSN initiative places explicit emphasis on ensuring that ICT enables new and more effective teaching methods, whereas the EEI's more basic objective is simply to enhance student familiarity and comfort with ICT. This is not a surprising revelation, that technology companies such as those mentioned earlier are less concerned with the pedagogy of teaching and learning but are more interested in getting 'their foot in the door.'

In many developing countries, the mobile phone has been playing an in-creasingly important role in business, and mobile phone manufacturers have been quick to promote their usefulness for educational purposes, as yet another con-duit through which people can access various forms of learning. Given young and not-so-young people's attachment to their mobile phones, it would be expected that they might be an extremely useful way of delivering education. However, the quality of education that could be provided is questionable, taking into con-sideration the shortcomings of mobile phones for this type of use, that is, costs of connection, strength and consistency of signal and small screen size. There are some interesting promotional videos on YouTube that have been uploaded by mobile phone manufacturers, demonstrating the usefulness of their products in an educational capacity. However, from viewing these videos, in some cases it would appear that only very prescriptive 'courses' of learning that can be reduced to brief question-and-answer sessions are on offer. If the learner answers correctly, he or she progresses to the next question; if incorrect, the student is then provided with the correct answer. There is very little real educational content apparent, as there is no explanation for either a correct or incorrect answer: This is merely learning by rote and no attention has been given to the pedagogy of elearning. In fact, comments made by users of mlearning confirm this; they say that they do not find learning in this way satisfactory, as it is difficult to take it seriously—it is more akin to playing a game.

In Africa, although elearning is being utilised, mlearning is gaining some ground, as it enables learning that is personal, portable and flexible, according to Traxler (2010). In his view, the difference between mobile learning in Africa and that in, for example, Europe, is that the former has different educational traditions and priorities. In Europe, the focus is more on personalisation, inclusion, participation and lifelong learning, whereas in Africa, it is a response to poor connectivity, mains electricity and the availability of computers. The spread of mobile phones and the 'vigour and talent' of mobile phone networks are also reasons for the take up of this technology. Traxler went on to explain that the challenges are for mlearning projects to become large-scale, sustained and sustainable, equitable, accessible and inclusive. His experiences so far have thrown up a number of questions, not the least of which is how to reduce one digital divide without creating or increasing others. This is the problem with ICTs; they can be costly to acquire, need technical knowledge to maintain, require some kind of network and need to be easily available and accessible. These requirements, if not met, will only exacerbate the digital divide.

Conclusion

It is abundantly clear from the aforementioned and other research highlighting similar problems that the answer does not lie in the use of new technologies in all circumstances but that we should be looking more at education, delivered by any means, for sustainable development. Governments and policy-makers need to look beyond the rhetoric of elearning and even mlearning and consider the best methods for educating their citizens, taking into account the local infrastructure and which mode(s) of delivery would be most appropriate in any given situation. Avgerou and Madon (2005), in their paper which discusses the information society, the digital divide and developing countries, stated that little attention has been paid to the purpose for which communities need knowledge and what knowledge is appropriate for particular groups. They argued that

> . . . the root of counter-development obstacles to ICT, that find their expression in terms of the digital divide problem, might be the extent to which the information society conveys aspirations, and privileges technologies, information and

knowledge that are irrelevant to the way the majority of people in many communities in developing countries live their lives. (p. 1)

There are numerous small-scale projects, especially in developing countries, that aim to educate communities that do not rely on the use of new media technologies. As mentioned in the opening paragraph, slates and blackboards are still widely in use in many areas of the developing world, and these may be the most appropriate, available and affordable technologies for that situation. The most important thing is to educate communities by any means and give them the requisite knowledge and tools to enable sustainable development in whatever part of the world they happen to live. The commodification of education as an enabler of the increasing privatisation and marketisation of educational provision through the widespread use of new technologies is clearly not leading to greater equality in either developed or developing countries but is only serving to further widen the digital divide. Technology by itself cannot create global equity—it is not a panacea for all the world's ills; this is a technologically deterministic view that is crediting technology with a power that it does not and cannot possess.

EIGHT

Arable Agriculture, Food, Technology Choice and Inequality

Peter Senker

This chapter is based on the proposition that a global food and farming system fit for purpose should ensure that the world's 7 billion population (expected to rise to 9 billion by 2050) has access to sufficient food to enable everyone to live healthily, without causing more than the minimum of damage to the environment and biodiversity.

It is suggested that the present system is dysfunctional. The overwhelming majority of choices of which technologies to develop and deploy are made centrally by large multinational corporations. Evidence is put forward that shows that an agricultural system based on local technology choices by small farmers, in cooperation with scientists, made in the light of local climatic and soil conditions, would likely be better, from several points of view. Such a system could ensure that the world's huge population had available to it more food more closely attuned to its dietary needs; it could protect the environment more effectively; and last, but by no means least, it could supply more and better employment to rural populations, thus reducing the rapidity of the drift of global populations away from the countryside to the cities.

At present, only 57.4 percent of the world's population consume a reasonable amount and quality of food to keep them in good health. About 28 percent consume too little food and about 14.7 percent consume too much. Economic growth and technological change have combined to lift hundreds of millions of people out of poverty and severe deprivation. But there are huge numbers of people suffering from severe deprivation—hunger, starvation and poor health: these include people who possess insufficient land on which to grow sufficient food for themselves and their families, together with poor and unemployed people who cannot afford to buy enough food. Hunger in terms of lack of access to a sufficient amount of the major macronutrients—carbohydrates, fats and proteins—afflicts about 925 million people. Perhaps another billion suffer from 'hidden hunger,' resulting from inadequate micronutrients such as vitamins and minerals. This brings risks of physical and mental impairment. In contrast, about a billion people each consume too much and suffer from chronic conditions such as type 2 diabetes and cardiovascular diseases (Foresight, 2011, pp. 9–10).

Over the past century, the total fixation of nitrogen globally has doubled. Increased use of nitrogen fertilizers has enabled massive increases in agricultural production, but it has had substantial adverse effects on the environment and human health. These effects include damage to water, air and soil quality and to biodiversity. There are widespread problems with soil loss due to erosion, loss of soil fertility and salination. Water extraction for irrigation exceeds replenishment in many places. Overfishing is widespread (Foresight, 2011, p. 40; Sutton et al., 2011). Meat consumption is rising fast, particularly in developing countries. Producing meat—especially from intensively produced grain-fed animals—is very expensive in terms of the resources required to produce a given amount of nutrition and in terms of environmental pollution (see Chapter 4 in this volume). This chapter concentrates on arable agriculture and food in developing countries. Food production and consumption also pose serious problems for developed countries, such as obesity, food waste and environmental degradation. But in developing countries, issues relating to agriculture and food are frequently a matter of life and death.

The US continues to exert powerful influences on world agricultural production and on food distribution and consumption. Accordingly, a brief history of US agriculture follows. The chapter goes on to consider food security and insecurity and their causes, along with prospects for food sovereignty. India provides an interesting case study, particularly in relation to the varying effects of the Green Revolution on regions of the same country with different conditions for agriculture, particularly in respect to water availability. Kerala provides an example of an attempt in one region to adopt policies different from those adopted in India in general, although of course, Kerala's freedom of action is severely constrained by national and international factors (Kannan, 2003, p. 3). Nevertheless, Kerala's experience is of interest because its radical reforms produced benefits to the state's people that no other Indian state and few other third world nations have accomplished. Increased agricultural production is vital but not sufficient to achieve poverty reduction and economic development. There is intense controversy about the future of agriculture and about which technologies are needed to meet the challenges.

US Agriculture

The US government, foundations and industry have become most powerful actors in the world's agricultural and food industries over the past one hundred and fifty years. There was abundant fertile land in the US in the nineteenth century. The federal government granted farmers large tracts of land free or cheaply. Topsoil in the Midwest was rich with nutrients derived from thousands of years of decaying plant matter. "The relatively mild climate and especially the large areas with frequent and reliable rain provided ideal conditions for grain" (Roberts, 2008, p. 17). US farms were huge and farm labour was scarce, creating incentives for farmers to adopt labour-saving technologies. Productivity increased rapidly, and Americans began to consume enormous quantities of food. Production outstripped demand, so US farmers looked to exports to dispose of rapidly rising production: "The emergence of an international food system built on railways, shipping routes and new preservation technologies spurred by free trade enabled suppliers in the United States together with suppliers in Australia and Argentina to prevent people in Europe from starving" (Roberts, 2008, pp. 17–19).

The US federal government developed a system of publicly funded farm programmes designed to increase output but also to protect farmers from harvest failures and market crashes. It set up universities to do research, provided extension services to train farmers in new technologies and provided land grants. Dams, irrigation channels and reclamation projects were undertaken to facilitate agriculture in dry regions. A massive rail network was constructed to transport produce from where it was grown to the big urban areas and to ports where it could be exported (Roberts, 2008, p. 20; Stiglitz, 2002, p. 21).

Traditional methods for replenishing soil fertility with manure and crop rotation were inadequate to replace the nutrients needed by the new fast growing crops. Dust bowls were created through the depletion and loss of physical integrity of soils, which the new crops caused. The Haber-Bosch process was utilised extensively to use nitrogen from the atmosphere as the basis for fertilizers (Roberts, 2008, pp. 19–21). Plants such as corn, which would grow larger but were also more uniform to facilitate harvesting mechanically, were developed. By the end of the twentieth century, the US was growing nearly half the world's soybeans and corn and exercised considerable control over world grain and food markets (Roberts, 2008, p. 25).

The US Cold War Strategy

After the Second World War, third world development became an integral part of the US government's Cold War strategy. Food, food aid and agricultural production played key roles. The Marshall Plan included distribution of food to hungry European populations. But by the early 1950s, European agricultural production had recovered. European farmers wanted US food aid to stop because it was preventing them from selling food locally. The focus of American food aid turned to developing countries—particularly to Asia—where several countries were perceived as moving toward communism. "The hungry might be rendered less trouble more grateful and . . . more dependent if provided with cheap food" (Patel, 2007, pp. 90–91). As markets for US food in Europe declined, in 1954 President Eisenhower signed Public Law 480. This made it possible for governments fighting trades unions or left wing political opposition to access the US strategic grain reserve (Patel, 2007, p. 91). During the Cold War, the US followed policies of promoting US international trade and overseas investment in cooperation with major US corporations and foundations, attempting to impose a US model of agricultural development on developing countries. In contrast to its domestic policies of supporting its own farmers, the US government made no efforts to protect developing country farmers from harvest failures and market crashes. Indeed, in cooperation with the IMF, it acted to discourage developing country governments from acting in this way (Stiglitz, 2002, p. 20).

The US government initiated a participant training programme that brought 6,000 foreign participants a year to the US for up to twelve months of study and training related to development in their own countries. Two important aspects of this were education (24 percent of participants) and agriculture (18 percent), both of which have continued in various guises ever since. By 1963, 80,000 participants from 80 countries had become ministers, senior civil servants, and so forth, in their own countries. US government goals included fostering a vigorous and expanding private sector in less developed countries. In 1964, a senior official stated that the US government's long-range political goals included "to insure that foreign private investment, particularly from the United States, is welcomed and well treated" (George, 1976, pp. 70–73).

Since the end of the Second World War, the General Agreement on Tariffs and Trades (GATT) had coordinated the lowering of tariffs on industrial goods among its members. From 1984, nearly eight years of negotiations led to the inclusion of agriculture in the remit of the World Trade Organization (WTO), which succeeded GATT in 1994. The US and Europe retained their huge subsidies to domestic agriculture, while developing countries agreed not to subsidise theirs (Stiglitz, 2002, p. 7).

Rapid expansion of food production occurred largely as a consequence of spectacular triumphs of US Cold War strategy, as illustrated by the case of India, discussed later, and in Indonesia. In Indonesia, President Sukarno started nationalisation and land reform. But interventions largely initiated by institutions in the US culminated in Sukarno's overthrow by military force, changing the government to one favourable to private enterprise, which led to successful implementation of the Green Revolution (George, 1976, pp. 79–83; Ransom, 1974).

The Green Revolution

After the mid-twentieth-century, a group of powerful agricultural technologies, in particular 'Green Revolution' technologies, were developed by scientists in international research centres primarily in the US, adapted in national research institutions, adopted by extension agencies and agro-chemical and seed companies and marketed to farmers. These technologies include uniform high-yield crops, mechanical and energy inputs and synthetic chemicals. They reduce indigenous biodiversity and cannot be used by small resource, poor subsistence farmers. In both India and Indonesia, the dissemination of the Green Revolution can be seen as arising from the success of the US's Cold War strategy. The Green Revolution benefited middle-income farmers in some developing countries (Senker, 2000).

The Green Revolution was started in Mexico in 1943 with the introduction of new, high-yielding dwarf varieties (HYVs) of wheat and corn. Similar developments were applied subsequently to rice. Between 1964 and1967, wheat output tripled and corn output doubled. Between 1965 and 1966 and 1972 and 1973, wheat and rice acreage planted with HYVs in developing countries increased dramatically (George, 1976, p. 145). India, Pakistan and Turkey followed Mexico, Taiwan, the Philippines, Sri Lanka and India in planting the new rice strains. This increased crop yields and aggregate food supplies. But it involved expansion of an industrial farming model, which could only be adopted by richer farmers who had access to adequate supplies of water and could afford expensive inputs of fertilizers and pesticides. It was mainly adopted by farmers to meet the needs of urban areas or export demands for food, rather than by poor rural farmers to feed their families. The Green Revolution discriminated against subsistence farmers and contributed to the loss of food self-sufficiency and agro-biodiversity at the local level among many areas of Asia, Latin America and Africa. Reliance on chemical fertilizers resulted in new 'ecological diseases' and also made developing countries' food production dependent on expensive imports of agro-chemicals and machinery (George, 1976, pp. 114–116).

Growing HYVs involved securing supplies of fertilizers and pesticides that generally have to be imported by developing countries from developed countries, principally the US. Growing these crops requires more frequent and more precise irrigation (George, 1976, pp. 115–116; Patel, 2007, p. 120). Farmers had to ensure that the necessary inputs were available. This excluded all but the largest farmers (George, 1976, p. 119). The Green Revolution contributed to global food security by increasing food supply, but it entrenched an unsustainable, inequitable food system that exacerbated environmental degradation. Initially, it increased the need for labour to spread fertilizers and pesticides and to gather in two harvests per year. But increasingly, machines were used by large farmers to reduce costs, thus reducing the demand for labour and the numbers of people able to benefit. For example, high pressure sprayers were used to deliver pesticides. Work previously done by women was increasingly handled by machines (Kropiwnicka, 2005).

The Green Revolution helped feed urban consumers by increasing marketable surpluses. But it did not help rain-fed farming, whose production may be adversely affected by droughts. As a result, income disparities between irrigated and rain-fed regions have worsened. The prospects for technological advances in rain-fed areas are hampered by limited and uncertain rains that often make water a critical constraint in plant growth and by the diversity of local growing conditions that limits the geographic applicability of improved technologies. This resulted in increased inequalities in living standards between farmers who could afford to buy the necessary inputs and those who could not (George, 1976, p. 124). Moreover, Green Revolution gains have tailed off. Salinisation of irrigated areas, pest increases, declining returns to input applications and water supply problems began to hit hard. In the midst of surplus, there are still people who cannot gain access to food because of lack of land ownership or access to land (Scoones, 2006, p. 26).

The Green Revolution was promoted by the US government, by foundations and multinational corporations, as an integral part of the US's Cold War strategies. It helped secure domination by US-based multinational corporations of a large proportion of developing country agriculture, including domination of the pattern of crops planted, supply of technology and inputs, in particular, fertilizers, pesticides and seeds (George, 1976, pp. 79–82).

Food Security and Sustainability

Food security is achieved when all people at all times have physical and economic access to sufficient, safe and nutritious food for a healthy and active life. Food insecurity is the absence of food security and applies to conditions such as famine and periodic hunger and uncertain or inadequate food supply. Undernourishment is when there is insufficient energy intake. Malnutrition is caused by deficiencies or imbalances in energy, protein and/or other nutrients. Deficiencies in micronutrients (vitamins and minerals) can also affect mental and physical health. The nutritional adequacy of food includes their physical ability to absorb nutrients. This can be affected by health factors such as intestinal parasites. Chronic hunger is a constant or recurrent lack of food and results in underweight and stunted children and high infant mortality. A famine can bring about deaths from diseases resulting from debilitation, from breakdown of sanitary arrangements that may lead to spread of infection or from population movements in harsh conditions (Sen, 1999, p. 169).

The components of food security include availability of land, access to food produced either locally or elsewhere, the capacity to produce, buy or acquire food and the stability of this access over time. These factors are affected by physical, economic, political and other conditions within communities. Entitlement to acquire food for a family can be acquired by farmers working on the land to grow produce or by people in employment acquiring sufficient income with which to buy food. In the short term, hunger and famine result from people suddenly experiencing deprivation of such capabilities with which to acquire food. Famine may result for groups of people who suddenly lose their entitlements to food even when there is plenty of food available in the areas in which they live (Sen, 1999, pp. 161–169).

In 2000, world leaders committed themselves to the Millennium Development Goals, which include the goal "to reduce by half the proportion of people who suffer from hunger between 1990 and 2015." (United Nations General Assembly, 2000). Seventy percent of the world's poor people, including the poorest of the poor, and 75 percent of the world's malnourished live in rural areas, where most are involved in agriculture. Yet rural poverty remains stubbornly high, even with rapid growth in the rest of the economy. Chronic hunger and global food security will remain an important concern for the next fifty years and beyond, as the world's population grows from its current 7 billion (2011) to more than 9 billion, most of whom will live in developing countries. Improved food security is important for global reduction of hunger and poverty, and for economic development.

By 2003, the proportion of world population that was undernourished had only decreased from 20 percent to 17 percent. In 2008, it was estimated that recent increases in food prices would add 100 million to the 850 million people then affected by hunger in developing countries. Since then, world food prices have increased sharply (Bretton Woods Project, 2008; Thompson et al., 2007).

Hunger, poverty and disease are interlinked. Hunger reduces natural defences against most diseases and is the main risk factor for illness worldwide. People living in poverty often cannot produce or buy enough food to eat and so are more susceptible to disease. Sick people are less able to work or produce food. Nutrition is an essential foundation for poverty alleviation and also for improving education, gender equality, child mortality, maternal health and disease. Hunger is a major constraint to a country's immediate and long-term economic, social and political development. Diseases such as HIV/AIDS and malaria affect food security by undermining people's capacity to produce food or to work to acquire the resources with which to buy it.

In addition to its health, economic and social impacts, disease also affects food security and nutrition. Adult labour is often reduced or removed entirely from affected households, and those households then have less capacity to produce or buy food, as assets are often depleted for medical and/or funeral costs (Parliamentary Office of Science and Technology, 2006; Thompson et al., 2007, p. 12).

Peter Senker

Famine can result when shocks such as natural disasters, armed conflict or drought affect vulnerable populations. Natural disasters and climate particularly affect those in countries that largely depend on rain-fed farming and those highly dependent on agriculture. Poor people are less able to cope with the impacts of climate shocks and variability. These events can result in massive crop losses, loss of stored food, damage to infrastructure and consequent increases in food prices. Armed conflict can cause food emergencies, reverse economic growth, destroy schools, roads and hospitals and force migration.

The weak bargaining position of food producers in developing countries contributes to food insecurity. A high proportion of food producers work in the informal sector. Few have basic social protection. Child and bonded labour are prevalent. Concentration in food production and distribution has been increasing. Huge agribusinesses buying from developing countries have gained bargaining power over their suppliers. Such buyers continue to pay relatively low prices even when international market prices increase, and they continue charging high prices to their customers when prices fall (Lines, 2008). Fair trade and the range of products it covers have been growing fast—now including coffee, bananas and other fruits, cocoa, tea, sugar and juices. It has benefited many developing country farmers: In 2008, it covered nearly $6 billion of trade. But it still covers less than 1 percent of world food trade (De Schutter, 2009, pp. 5–6, 18–19).

Risks of food insecurity are increased by low rates of agricultural production. Such risks may be increased in a poor economy. Growth in output in general tends to protect against famine insofar as if food is not being produced, it can be bought. Recently food production, especially in many developing countries, has failed to keep pace with rapidly growing demand. Over the past twenty years, external support for investment in agricultural productivity has fallen rapidly. In contrast, agricultural subsidies in rich countries (primarily the US and the European Union) now run at around US $1 billion per day. This is more than six times rich countries' entire aid budgets, despite WTO agreements aiming to increase international trade through a reduction of trade barriers (Bretton Woods Project, 2008).

While subsidised imported food may help prevent famine in the short term, it depresses food production in developing countries when farmers cannot afford to compete with it. Sudden increases in the price of imported food or sudden decreases in its availability can trigger famine in developing countries. Democratic governments are strongly motivated to prevent famines by the prospects of losing power if they fail in this, while the motivations of authoritarian governments to reduce famine are far less (Sen, 1999, pp. 179–181). In the years 2008 to 2010, sharp price increases have been caused by higher energy and fertilizer costs linked to increasing oil prices, and the impact of biofuels has also been significant. Social protection safety nets can be used to increase poor people's access to food—by transferring food or goods to poor people or through temporary employment creation. Increased agricultural production is vital, *but not sufficient,* for poverty reduction and economic development. Science and technology can help improve food security through an increase in food production (using new crop types, etc.); improvements in cost and quality of food storage, processing, packaging and marketing; labour saving technologies; and better communications (Parliamentary Office of Science and Technology, 2006; Sen, 1999, pp. 160, 168, 204; Thompson et al., 2007).

India—A Case Study

Famine in 1943, particularly in Bengal where 3 million people died, led to the appointment of the Food Grains Policy Committee, which recommended procurement of food grains from surplus areas, rationing to achieve equitable distribution and statutory price control. The Public Distribution System (PDS), which distributes food grains through Fair Price Shops (FPS), was expanded considerably (Patel, 2007, p. 128). Controls became pervasive during 1944 to 1947, as food shortages continued. A further Food Grains Policy Committee, appointed in 1947 after independence, recommended gradual abolition of food controls and rationing and imports to maintain central reserves to be available in case of crop failure. In 1950, the introduction of National Planning provided for the state to take responsibility for nutrition and public health. The government undertook to plan "simultaneously for increased production of wealth and for more equitable distribution of the wealth produced" (Singh, 2006, p. 20). Provision of food to vulnerable sections of the community at reasonable prices became an important objective.

During 1953 to 1955, decontrol was tried, but it failed in 1956, when food prices rose rapidly. Several control measures were reintroduced. In 1954, the US government had enacted Public Law 480, which legislated for food aid to developing countries. In August 1956, the US and Indian governments agreed that India would import 3.1 million tonnes of wheat and 0.19 million tonnes of rice during the next three years. This was the beginning of an Indian policy of relying on continued imports of food grains and their distribution through a system of Fair Price Shops in large quantities and at low prices, which lasted for about ten years. But in August 1963, prices began to rise (Singh, 2006, pp. 19–21).

Indian farmers could not compete with imports of cheap subsidised wheat from the US. Indian wheat production was static. Many poor, rural Indians were on the brink of starvation. India's Prime Minister Nehru's observation that China had boosted agricultural production substantially through land reform encouraged him to consider this for India. As in many developing countries, a few rich people own most of the land in India. It is possible to carry out land reform peacefully and legally and it can be accompanied by measures to extend access to credit and extension services that train farmers in new techniques. Indeed, subsequently, land reform contributed to the welfare of several countries such as Korea and Taiwan (Stiglitz, 2002, p. 81). Nehru suggested that cooperative land management between landowners and those working the land could increase yields, but his proposals were opposed by landowners' representatives in the Indian Parliament (Patel, 2007, pp. 121–122).

The rains failed in 1965, and drought led to food shortages: There were particularly bad harvests in Bihar and Gujarat. The famine was more a problem of food distribution than availability, so grain-poor states were linked with grain-rich ones to try to alleviate the problems of those grain-poor states. This was not entirely successful. Nehru died in 1964 and was succeeded by Lal Bahadur Shastri. As a consequence of food shortages in 1963 to 1965, there were rural uprisings by peasants calling for land reforms. Shastri expressed disapproval of US bombing in Vietnam, which President Johnson viewed as an expression of pro-Communist sympathies. Accordingly, Johnson did not immediately renew India's Public Law 480 food contract and shifted it from an annual to a month-to-month basis. But provided that Shastri abandoned land reform and that India became more sympathetic to US policy in Asia, the US offered to resume food aid on a long-term basis and would offer India new agricultural technologies that would alleviate its food supply difficulties without involving rich landowners losing land (Patel, 2007, pp. 120–122).

Shastri died in January 1966 and was replaced by Indira Gandhi. In March, she travelled to the US and established a good personal relationship with President Johnson. The failure of the rains again in 1966, together with grain hoarding and poor distribution, worsened the threat of famine. Indira Gandhi abandoned her father's nonalignment policies and initiated the Green Revolution in India (Adams & Whitehead, 1997, pp. 198–208). Food aid came to an end in the 1970s. Meanwhile, the PDS offered prices to farmers that encouraged them to sustain the supply of food grains. Until the 1970s, the PDS was significant mainly in urban areas. During the 1980s, its coverage extended to rural areas, first in the south and later throughout India. In the severe 1987 drought, public food stocks and redistribution contained famine (Patel, 2007, p. 129). However, it became apparent in the 1990s that the PDS had become very inefficient in terms of helping the poorest households: They received little subsidised grain, but PDS may have acted to increase the price of open market grain, which they had to rely on (Singh, 2006, pp. 8–10). The PDS system was revised in 1992 and 1997. The grain distributed by the government fell from 17.2 million tonnes in 1997 to 13.2 million tonnes in 2001. However, while the PDS system needed reform, the reforms probably went too far toward dismantling the system. In some states, including Kerala, PDS may have played a useful role in providing poor people with much-needed food supplies (Patel, 2007, p. 129).

Vast state expenditures on the Green Revolution were concentrated in the Punjab, the most fertile area of the country. Punjab is a state with 24 million people, about 2 percent of the population, and produces over 12 percent of India's food. Food grain production there increased from about 3 million tons in 1965 to 1966 to over 25 million tonnes in 1999 to 2000. But three quarters of Indian farmers living in poorer states without access to large areas of land were marginalised and failed to benefit from the Green Revolution. Even in the Punjab, the number of smallholdings dropped sharply because many farmers could not afford the necessary irrigation or fertilizers (Patel, 2007, pp. 125–126).

Thus, as Scoones (2006) suggests, "The focus on equity, social justice and community development during the 1950s and early 1960s gave way to an emphasis on technology-led transformation of agriculture" (p. 24). And furthermore, "the implication that the Green Revolution was freely and willingly adopted by the government of India, let alone its people, is unwarranted. On the contrary, a good number of peasant movements were fighting for alternative ways of doing things" (Patel, 2007, p. 124).

The impacts of the Green Revolution in India were uneven. The release of a series of dwarf wheat varieties followed by rice varieties boosted production massively, initially in the irrigated plains but later more widely. Total production of wheat increased seven times between 1960 and 1961 and 2000 and 2001 and of rice seven times. Much of this was achieved through yield increases, which doubled in rice and tripled in wheat (Choong, K. Y., n.d.). The Green Revolution helped feed people in India's rapidly growing cities, but in the last quarter of the twentieth century, the poorest 30 percent of the Indian population gained little improvement in nutrition (Thompson et al., 2007). The principal aim of Indian biotechnology now is to produce high quality products with a world market, and this militates against the development of products that meet the diverse needs of local farmers in a variety of climatic and soil conditions (Scoones, 2006, pp. 39, 354).

Kerala

Since 1957, the Indian state of Kerala has adopted political solutions to agricultural and social problems rather than adopting technological fixes such as the Green Revolution. The Keralan government's policies of land redistribution, food distribution, employment guarantees, health care and education resulted in the highest literacy, health and general welfare in India. Indeed, despite the poverty of Kerala's population, literacy levels and life expectancy are higher on average than in some parts of the US. Moreover, the Keralan approach has provided more enduring benefits than the Green Revolution. In the 1990s, twenty years after the Green Revolution, while malnutrition seems to have increased almost everywhere else in India, indicators of health and welfare remain high in Kerala (Kannan & Vijayamohanan Pillai, 2004; Patel, 2007, pp. 126–127).

Reforms were sustained despite continuing opposition from the Indian central government. In the late 1950s "There was horror (particularly in the United States) when Kerala went Communist that it would start an infection that would ravage the whole of India like a plague" (Adams & Whitehead, 1997, p. 171). The Congress Party received money from the US Central Intelligence Agency to help it oust the Communists in Kerala (p. 172).

Kerala's radical reforms produced benefits to the state's people that no other Indian state and few other third world nations accomplished. Land reform removed the threat of eviction of tenants, making possible their greater political participation without fear. Breaking the hold of the landed elite over productive resources liberated many small farmers from the effects of highly unequal land ownership. The Kerala Agricultural Workers' Act protects farm workers through laws covering wages and working conditions and has served to increase the incomes of many poor farmers. Kerala's school and nursery feeding programmes ensure at least a minimum of food to nearly everyone in the state. By offering low prices, shops help reduce the plight of the most potentially undernourished groups and help them remain free of private moneylenders whose practices are a source of exploitation and misery in many third world rural areas. Kerala has much lower per capita income than China but overtook China in terms of life expectancy and infant mortality (Sen, 2009). But Kerala's reforms failed to raise agricultural production. Moreover, the generally low level of economic development results in a relatively high level of unemployment, especially amongst well-educated sections of the population. It would, of course, be better to have both good health and education and a higher income (Sen, 1999, p. 48).

Competing Visions of Innovation

There is intense controversy about the future of agriculture and about which technologies are needed to meet the challenges. For example, the successes of the Green Revolution in India have tailed off and agriculture is no longer providing growing income or employment opportunities to sustain livelihoods. Advocates of biotechnology argue that its application can help increase production, reduce costs and improve product quality. They suggest that biotechnology could have major impacts on reducing poverty, boosting incomes and employment opportunities in poor, rural areas of the country. Others argue that the public sector lacks the capacity to supply new technologies and that poor producers could not afford to invest in the expensive inputs that they would inevitably require. Unless there are radical changes in the ways in which it is used, the adoption of biotechnologies could lead to large numbers of people losing their land, accompanied by rural destitution and distress. The likely outcome may well include acceleration of current trends to larger farms, accompanied by rural destitution and increased poverty: The policies of the government of Karnataka seem to be leading in this direction (Scoones, 2006, pp. 40–41).

The dominant vision politically is of a 'modern' agriculture from the Green Revolution to the current Gene Revolution as a standard, preferred pathway to development. Such a perspective centres on technology, production and growth. Key elements of the modern agrifood 'system' involve a wide array of external expensive inputs such as research and development, fertilisers, seeds and irrigation. Technology-driven economic growth through sustained innovation and trade is envisaged as providing pathways out of agriculture or a shift of subsistence-oriented 'old' agriculture to a modern, commercial, 'new' form of agriculture, with wider poverty reduction aims achieved through trickle-down and employment benefits from improved agriculture-led growth (Thompson et al., 2007). This vision perceives the role of agriculture as an 'engine of economic growth' and sees the economic and social transformation of the agrarian economy—from backward to modern, from subsistence to market-orientated—from 'old' to 'new' agriculture.

The development of agricultural biotechnology has been driven principally by commercial interests and has resulted mainly in standard solutions that involve expensive external inputs and reductions in crop diversity. Like the Green Revolution, it seems likely that the 'Gene Revolution' will be delivered in relatively expensive packages that risk amplifying inequalities further. During the 1990s, some major multinational corporations claimed that genetic engineering would increase the productivity achieved by farmers in developing countries and alleviate poverty and hunger. But research has been concentrated in areas where it was thought likely to offer big markets in developed countries, rather than in those that would benefit developing countries such as drought-resistant crops for marginal lands or foods that have a high nutritional value. Few of the foods produced so far or being researched and developed are foods that the hungry can afford. Biotechnology companies have concentrated on a restricted range of patented transgenic crops designed for capital-intensive production for large markets (Senker, 2000).

WTO rules prevent farmers from reproducing patented seeds (Altieri, 2005). Efforts by a US company to patent basmati rice caused an outcry and highlighted the dangers involved in patenting. Intellectual property protection to crops will have negative consequences for poorer farmers (Commission on Intellectual Property Rights, 2002; Senker & Chataway, 2009). The success of biotechnology related PPPs has been insignificant so far. Ayele and Wield (2005) conclude that

PPPs contribute to science and technology capacity building but do not always involve producers and users at an early stage. This limits their effectiveness for poorer farmers. High levels of inputs such as fertilisers and pesticides, together with the reduction of biodiversity that tends to result, are liable to have adverse effects on the environment.

Alternative visions centre on food sovereignty, "a vision that aims to redress the abuse of the powerless, wherever in the food system that abuse may happen" (Patel, 2007, p. 302). Such participatory visions emphasise working with natural systems, generating improved livelihoods with more ecologically attuned production systems, with the empowerment of local farmers being seen as central to achieving both economic and ecological sustainability (Thompson et al., 2007, p. 43). It could be important to try to direct agricultural research to serve small farmers better. Technology and societal choice could be more closely entwined. To be effective in developing countries, agricultural innovation needs to take place on the basis of local and traditional knowledge, cultural preference and local environmental conditions. This could involve participation of local farmers in technology choice and the development and application of a wider range of technologies requiring low inputs adapted to varying needs and contexts. The use of some biotechnology, some organic farming all directed at benefiting the poor, could be involved in the implementation in Swaminathan's (2000) vision of an 'evergreen revolution' (Scoones, 2006, pp. 39–43).

New Horizons—Food Sovereignty or Corporate Control?

About 2.5 billion people live off the land worldwide. Grassroots movements aimed at food sovereignty are growing rapidly in size and strength in many countries (Branford, 2011). Inequality is most harmful to those people whom it deprives of the most basic necessities of sustaining a healthy life. Prominent in this group of about a billion people currently in developing countries are those whom poverty and landlessness deprive of the basic necessities for pursuing a healthy life—those who lack sufficient food and clean water to eliminate hunger, starvation and disease. Technological advances such as the Green Revolution and biotechnology, even supplemented by massive aid to developing countries, are inadequate as solutions.

Those who possess insufficient land on which to grow enough food for themselves and their families, together with those who are unemployed and cannot afford to buy food, comprise the vast majority of those who suffer from severe deprivation—hunger, starvation and poor health. Economic growth and technological change have combined to lift hundreds of millions of people out of poverty and severe deprivation. But the present economic system, driven by the pursuit of profits, continually adds to the numbers of people suffering from poverty land-lessness, unemployment and severe deprivation.

The world food and agricultural system is profoundly dysfunctional in terms of its effects on the interests of the world's people, especially the poor, and also in terms of its environmental impact. The emphasis of world agricultural pro-duction needs to shift toward food sovereignty, including the preservation and technological development of local low-input systems, in which world food needs could be met to the great advantage of billions of small farmers and with far less environmental damage.

PART III

Technology, Opportunities, Threats and Contested Futures

In this section, authors explore issues relating to the opportunities offered by new technologies and the often very contradictory nature of their development and the uses to which they are put. It is widely recognised that technologies are a catalyst for change but their potentials can be shaped in many ways and by a rich mix of social, cultural, political and economic factors. In some instances, the radical potentials of technologies may be utilised proactively to challenge existing structures of inequality. This may be the case particularly in the early days of technological development when technologies may be appropriated in ways that are entirely unpredictable and even undesirable to their original producers. In others, the economic and political processes that more often determine the nature of their development limit these radical potentials and effectively constrain our contribution to, and control over, their development and impacts on our lives.

In Chapter 9, Kathy Walker explores current policies for the commercial allocation and auctioning of the radio spectrum and the issues they raise. The notion of the radio spectrum as a 'public good,' a commodity held in common for the benefit of all members of society, is increasingly under pressure. The radio spectrum is still a scarce resource even though digital technologies mean that the spectrum can be used more efficiently and for more applications. This chapter explores the inequality of access to the policy-making process in relation to spectrum allocation that favours large, corporate interests and the commercial applications of the mobile phone industry whilst at the same time failing to make adequate provision for existing public service applications that have a value above and beyond their notional spectrum value. It considers potential new uses and applications that may be marginalised by more market allocation of spectrum and 'spectrum trading' and questions the implications of allowing market forces to shape the allocation of precious, public radio spectrum.

In Chapter 10, Charlotte Chadderton explores the contentious use and application of CCTV technology for surveillance purposes in schools in the context of debates about citizenship, social justice and participatory democracy. The use of CCTV is frequently justified on the grounds of the enhanced security and safety such systems can provide for society at large. However, Chadderton argues that the way this technology is used in schools, and the political context in which its use is promoted, both reinforces and reproduces social inequalities that contribute to the creation of disadvantaged populations on the margins of society.

In Chapter 11, Maxine Newlands focuses on the radical use of new media technologies to challenge the domination of the news agenda by traditional media. Although her research demonstrates the ways new social media enable protests to bypass mainstream media practices and facilitate the production of media by activists themselves, she argues that caution needs to be applied in assessing its overall impact and influence. The use of social media sites can suggest the protest has greater strength and support than might exist in reality and could potentially lead to a 'digital divide' amongst communities of potential protesters.

NINE

Invisible Medium, Virtual Commodity

Changing Perspectives on the Radio Spectrum: From Public Good to Private Gain?

Kathy Walker

> Here's the next big thing, people: trading bandwidth . . . . If you're not using your bandwidth capacity, we could sell it on. It's tradeable. But people don't think in those terms because it's a virtual commodity. Well, Enron *gets* virtual. We're changing business, we're changing people's lives, we're changing the world. (Prebble, 2009, p. 56)

When one of the main protagonists in the stage play *Enron*[*] delivers this exuberant speech, it meets a sceptical response—"trading bandwith?"—at the time an unexpected and improbable trading venture to the uninitiated and indicative of a continuously expanding and rapacious market-driven corporation ever on the lookout for new commercial opportunities. In the real world, however, by the start of the new millennium, bandwidth exchange that allows companies to buy and sell, that is, trade spare capacity on fixed telecommunication networks, had already taken off in the US fibre optics market. Enron Communications was a major energy company turned supplier of network bandwidth capacity and had begun to focus its attention on possible developments in radio spectrum exchange. RateXChange, one of a small number of online telecom trading exchanges that included Enron Communications, had initiated real-time (live) trading on its bandwidth exchange service in 2000 and was developing plans to set up the first ever secondary market in the trading of 'wireless' or radio spectrum, regulation permitting (Sapp & Jones, 2000). The air of incredulity conveyed in the fictitious exchange about bandwidth reflects a similar response decades earlier to a suggestion from the esteemed economist and Noble Prize winner Ronald

---

[*] The Author's Note to the play indicates that although the play is inspired by the real events leading up to the Enron collapse, it should not be seen as an exact representation of events.

Coase, that opening up the 'airwaves' or radio spectrum to the market and private ownership would be a way of improving efficiency in spectrum allocation. When Coase addressed the US Federal Communications Commission (FCC) in 1959 to present his argument for greater market allocation of radio spectrum rights, the FCC's first question was apparently, "Is this all a big joke?" (Faulhaber, 2005; Hazlett, 2001a).

Historically, the radio spectrum has been 'allocated' by governments for defence, emergency services, telecommunications and broadcasting, and regulators such as the FCC in the US and the Radio Authority* in the UK, had responsibility for managing the spectrum on behalf of the government. Allocation of spectrum for these applications was largely awarded by licences that conveyed rights of usage but not ownership. This was essentially because, as Smith (1985) points out, "the radio wave portion of the electromagnetic spectrum is a natural resource which, although not depleted through use, is of limited size" (p. 76). The notion that the custody and control of a finite medium such as the radio spectrum could or should be left predominately to market forces clashed fundamentally with the traditional recognition of the radio spectrum as a 'public good,' held in trust by the state and administered for the benefit of all society and its citizens. However, in the last decade, this regulated approach to the radio spectrum and its management has come under increasing pressure in international policy-making circles as a result of what Arnbak (1997) has called a 'mounting critique' of the "purely administrative international and national approaches to allocation and assignment of frequencies to users" (p. 3).

Extensive development in wireless technologies and applications has resulted in a growing demand for deregulation and liberalisation of the radio spectrum to allow wider and more easily accessible usage for new businesses and commercial initiatives. Governments internationally have been lobbied by corporate interests to 'reform,' as it is invariably termed, spectrum management systems and release greater tranches of radio spectrum for private development. National regulatory authorities (NRA) around the world have responded to these pressures to make new slices of the airwaves available by selling more and more spectrum, increasingly via auctions, and by facilitating increasingly liberalised approaches to spectrum trading. In 2000, for example, the UK government auctioned a newly released block of spectrum for third generation (3G) mobile phones, and to the astonishment of industry, academic and government observers at the time, bidders paid £22.5 billion for these spectrum licenses (Binmore & Klemperer, 2002). Following this very successful auction, the UK government has been preparing the bidding process for the terrestrial television spectrum that was finally released as a result of the switch-off of the analogue television signal in October 2012. This is termed the 'Digital Dividend,' and the 'sweetspot' in the radio spectrum has

* Superseded by the Office of Communications (Ofcom) in 2003.

generated enormous interest in terms of its potential allocation and the substantial revenue its sale is expected to generate for the government. Most recently, the sale of equivalent digital dividend spectrum in the US in 2008 netted $19.5 billion (BBC, 2008). In debates that are couched largely in economic terms, notions of equality and communality inherent in the concept of public goods have, for the most part, been replaced by the alternative philosophy of the commodification of the airwaves and the widespread commercialisation of its applications.

This commercial allocation and auctioning of the radio spectrum raise a number of concerns in relation to the policy-making process and its outcomes for society as a whole. The radio spectrum is still a scarce resource, even though digital technologies mean that the spectrum will be able to be used more efficiently. There may be many new uses and applications for this spectrum, probably some that have not even been thought of yet, which may be constrained by auctioning and trading spectrum in a purely economic fashion. Even the report co-written by Coase, titled 'Problems of Radio Frequency Allocation,' acknowledged that "five years ago no one foresaw the kinds of demand now being made for allocation of frequency spectrum on an international basis for space purpose. Even today the frequency bands that will be wanted for the operation of communication systems five years hence is a matter of great conjecture" (Coase, Meckling, & Minasian, 1995, p. 3) Similarly, different services have different needs and broadcasting is not just another communication technology, like mobile telephony. It also provides content and has a value above and beyond its notional spectrum value that needs recognition. In recent years policy debates and discussion in the UK have focused on arguments for less planned allocation of spectrum and more market allocation. 'Spectrum trading,' which allows successful bidders to sell on their licences without the subsequent involvement of government or regulator, technology and service neutral conditions, and licences of indefinite duration, have become an accepted part of Ofcom's practice in relation to spectrum allocations. This chapter examines the changes that have taken place in attitudes to, and policy for, the management of the radio spectrum and the pressures that have shaped these changes in approach. It explores the concept of the radio spectrum as a 'public good' and the issues of inequality that arise as a result of the increasing application of market economics to spectrum management. It considers current UK government policy in this area and the possible consequences of allowing market forces to dominate discussion of the allocation of this precious, public resource.

The Radio Spectrum and the Concept of a 'Public Good'

Very few of us ever think much about the radio spectrum; it exists, but we cannot see it, smell it or touch it. It is invisible but not undetectable for, as science has demonstrated, the radio spectrum, home to communication applications, is a relatively small part of this electromagnetic spectrum. This includes the range of all electromagnetic radiation from electric power at the long-wave end, to gamma radiation at the short-wave end and infrared, ultraviolet, x-rays and radio waves in between. The radio spectrum is divided into frequency 'bands' and current applications range from radio and television broadcasting, two-way radios, remote controls, radar (radio detection and ranging), cordless microphones (for theatre and concert use) and satellite communications, to more recent innovations such as mobile phones and Wi-Fi (broadband access technology). So, the radio spectrum is invisible, but it is becoming increasingly important to the way we live our lives and to the essential network of communication structures that underpin them.

The more information a signal has to carry, the more bandwidth it requires, so different applications have traditionally been allocated to different areas of the spectrum that suit their technical requirements and to protect them from interference. The radio spectrum, like many natural phenomena, is a finite entity, limited in capacity, but new digital technologies and compression techniques means less bandwidth is required for digital signals than analogue signals. As a result, more radio spectrum is becoming available for new and different technologies and applications. At the same time, however, the proliferation of new wireless communication applications and the growth in wireless internet browsing and ultrafast broadband have generated increasing demands for what is still limited spectrum space.

It is the finite nature of the radio spectrum and the fact that it is essential for all wireless communication that has contributed to the common perception of the radio spectrum as a public good. This is partly due to its historical development, when the use of radio frequencies was associated with a few essential services such as military communications and maritime navigation that "could not be provided through the private market," and therefore, "the only practical way of rendering such risky wireless services was by collective action" (Arnbak,1997, p. 5). In economic terms the features of a public good are usually defined in contrast to those characterising private goods. As Batina and Ihori (2005) explain, "a pure public good exists 'where the marginal cost of providing another agent with the good is zero, and where no one can be excluded from enjoying its benefits'" (p. 2). Public goods are therefore described as both 'non-excludable,' in the sense that they cannot be withheld from those who do not pay for them, and 'non-rival' in consumption terms, meaning that they may be consumed without reducing the amount available for others.

In the case of the radio spectrum, for instance, pirate radio stations have broadcast in the UK since the 1960s without the benefit of a licence and, as long as they have appropriate transmitting equipment, they can broadcast until detection and prosecution prevent them from doing so. Similarly, although a licence fee is required to receive broadcast signals in some countries such as the UK, it has not been a physical impediment to those television owners who choose to watch broadcast television in contravention of this requirement. Technically, it has been difficult in the past to limit the transmission and reception of these frequencies without a regulatory structure in place to restrict access by legal means. However, changes in technology that allow the encryption of broadcast signals have now made it possible to exclude access to signal transmission, allowing for greater commercial development. The radio spectrum is non-rival in the sense that it is not necessarily depleted by the number of users receiving signals or transmissions at any one time. It is regulation that has usually been employed to limit the number of authorised users of particular frequencies and to protect the integrity of the respective radio signals from interference from unauthorised users.

It is the features of non-rivalry and non-excludability and the collective nature of the use of public goods such as the radio spectrum that have generally meant that national governments have stepped in to facilitate the coordinated action needed for their development and to ensure equitable benefits for all. The features of non-rivalry and non-excludability are often used to explain government involvement in the organisation and provision of public goods and services, since they imply that there is very little financial incentive for private providers to supply them on a commercial basis. However, the collective nature of their use also underpins the recognition of them as resources held in 'common' and the notion that their development and use should correspond to public needs and interests rather than just for the benefit of narrow sectional interests or for the provision of purely market applications. The decision by governments to intervene in the regulation of public goods is therefore based not only on economic or technical grounds but generally underpinned by social and cultural responsibilities to society at large. Arnbak (1997) argues that it has been this 'trustee' model that has traditionally underpinned governments' policy for the radio spectrum and the systems of regulation and spectrum management they have implemented. Although governments or government-appointed regulators may 'manage' the radio spectrum and other common resources, they only hold them in trust for the public as well as future generations.

Over the past decade, significant developments in wireless technology have begun to undermine and redefine the distinction between *public goods* and private goods and have, at the same time, been used to challenge the function of regulators to preserve an equal balance between public and commercial access to the radio spectrum. This chapter examines some of the aspects of the regulatory systems for the radio spectrum in the US and the UK before moving on to explore the pressures for change that have caused a major rethink in government attitude and policy to spectrum management and allocation.

The Radio Spectrum and Deregulation

As suggested in the introduction, although the US is generally considered to be the home of the free market and commercial enterprise, the notion of applying ownership rights or market mechanisms to the radio spectrum had long been resisted. The 1934 Communications Act established the basic framework for the regulation of wireless communication by establishing an FCC that centralized telecommunications regulation and policy-making with a central guideline that it should act for the "public interest, convenience and/or necessity" (Smith, 1985, p. 288). It was given a wide range of discretionary powers by the US Congress, which included both common carrier provisions and provisions relating to broadcasting (although this did not include regulation of broadcast content). Although the US telecommunications system is privately operated, significantly, the 1934 Communications Act made it clear that "users of the electromagnetic spectrum had no property right to a license to operate, but were given a temporary license for use of a public resource" (Smith, 1985, p. 288). Hazlett et al. point out that, "licenses were assigned by comparative hearings ('beauty contests') on the grounds that chaos would reign in the airwaves were the rights to be sold" (Hazlett, Porter, & Smith, 2009, p. 2).

Ronald Coase's arguments for the reform of regulation relating to the allocation and management of the airwaves presented in his 1959 paper had included the contention that, contrary to popular opinion, the airwaves could be defined as property by separating the creation of resource rights from the assignment of those rights. This, together with his proposal for spectrum auctions (competitive bidding), was roundly rejected by the FCC and a range of other experts at the time. Until the early 1990s, the FCC relied on comparative hearings and lotteries to select a licensee from mutually exclusive applicants. However, by the early 1990s, the position on auctions as a means of allocating spectrum was already under review, and in 1993, Congress granted the FCC auction authority. The first such auction was held in 1994, when the FCC gave approval for licence sales to commence by competitive bidding, and by 2009 73 US auctions had been held with "27,484 licenses sold, and $52.6 billion paid" (Hazlett, Porter, & Smith, 2009, p. 1). Although these auctions introduced considerable levels of competition and market pressure to the primary allocation process, critics (2009) argued that they still formed part of a tight regulatory structure that fell far short of the property rights regime advocated by Coase. Therefore, although they generated considerable revenues for the US government, licence auctions on their own did not deliver the market allocation of spectrum or spectrum trading, which would allow this spectrum to be resold in the open market, and which was demanded by some proponents of deregulation. At the time of the Enron debacle, pressure for further deregulation was already building from a number of sectors including the new mobile communication and network technology industries, those involved in the new commodities markets and economists anxious to reform what they perceived as the economic impediment of the under-utilization of the radio spectrum. Hazlett (2001b) argued at the time that the US regulatory structure for the radio spectrum had set the stage for decades of anti-competitive policies, since "upstart wireless competitors must queue up at the FCC to gain a spectrum allocation via a formal Rule Making, as there is no private band owner who can sell, lease or rent airwaves to deliver these new services to customers"

By the beginning of 2000, the FCC appeared to have acknowledged these pressures, by announcing publicly that "they wanted to move towards a 'radical overhaul' of FCC policies to make wireless bandwidth markets possible" (Hazlett, 2001b). In 2003 the US government took the first steps in allowing companies to lease and trade radio spectrum licenses when the FCC made it easier for wireless companies to gain access to spectrum licenses held by others. As *The New York Times* reported, "the move followed heavy lobbying by the largest wireless carriers, including AT&T, Verizon and Cingular, as well as players on Wall Street like Cantor Fitzgerald that are hoping to serve as brokers or clearinghouses in the creation of a secondary market for swapping licenses" (Labaton, 2003). Subsequent amendments to FCC rules have led to a substantial volume of spectrum trading, with approximately "1,000 assignments to use spectrum traded annually, many through private organisations (band managers) authorized to grant usage rights and define interference limits" (Wellenius & Neto, 2007, p. 17).

In the UK and Europe, there has been a similar transformation in attitudes toward the airwaves, underpinning a comparable pattern of liberalisation and deregulation of systems for the radio spectrum management. In the UK, this follows along pattern of deregulation and liberalisation of the telecoms market, beginning in 1984 with the privatisation of British Telecom, the publically owned provider of telecoms services, and the introduction of competition into the sector. In the 1990s, following the 1990 Broadcasting Act, the ITV/C3 regional broadcasting licences were auctioned for the first time on the basis of financial bids, albeit with a quality threshold requirement for broadcasting content, and the 1996 Broadcasting Act established a similar bidding process for the Digital Terrestrial Television (DTT) multiplexes. However, these licences were still awarded for a finite period, and in each case, they set out detailed conditions and legislative requirements about the service and content covered by the licence. The most significant step in the move toward deregulation of the radio spectrum in the UK came in 2000 when the Labour government set up the first 3G auction of 100MHz bandwidth. Auctioning mobile phone airwave licences to the highest bidder paved the way for the development of the 3G mobile phone networks that have enabled enhanced services including high-speed internet access and video content. The £22.5 billion raised at auction for the treasury occurred at the height of the telecoms bubble, which is generally believed to have driven up the bidding war. A report for the European Commission in 2004 noted that the UK was one of first countries in Europe to use spectrum auctions as a means of allowing the market to determine 'primary assignment' where spectrum usage rights (licences) were scarce (Analysis Consulting and Partners, 2004).

Following the financial success of the 2000 auction for the UK government and the publication of Cave's (2002) 'Review of Radio Spectrum Management,' it became clear that the UK government favoured further deregulation of the system. The report for the EU identified two distinct policies relating to spectrum trading: "trading, the transfer of spectrum rights between parties in the secondary market, and liberalisation, relaxation of restrictions on services and technologies associated with usage rights" (Analysis Consulting and Partners, 2004, p. 9). Tessa Jowell's statement when presenting the draft Communications Bill to Parliament in 2002 made it clear that spectrum trading would form a key strand of the Labour government's future communications policy, suggesting that more spectrum would be made available from existing capacity, and regulations on access would be liberalised:

> We will also extend the principles of deregulation and market competition to the allocation of the radio spectrum by introducing spectrum trading. Spectrum is to the modern age what iron and steel were to the first industrial revolution. It must be used efficiently. Companies need to know that they can gain access to spectrum so that they can bring their ideas to the market. In future, as well as being able to apply for a licence, firms will also be able to buy spectrum from an existing user. (Jowell, 2002)

One of Ofcom's[*] first initiatives was, therefore, to establish a consultation exercise on spectrum trading and to set out a timetable to bring this about. This initiative was a key element of Ofcom's programme of market-based 'reform' of the telecommunications sector, and as the new regulator said at the time, "if we introduce this system, it will mean that the UK will be the first country in the European Union to allow trading in licences to use the spectrum" (Ofcom, 2003, p. 3). Following the completion of the consultation process and the publication of the 'Spectrum Framework Review' in 2004, Ofcom went ahead with the introduction of this policy, explicitly designed to enable "licensees to buy and sell spectrum in the market ('spectrum trading') and reducing or removing unnecessary restrictions and constraints on spectrum use ('spectrum liberalisation')" (Ofcom, 2004, p. 1). Whilst the measures met with approval within the telecommunications sector, many of whom had supported the consultation paper's proposals, reports of this change of attitude to the management of the radio spectrum did not underestimate their significance. *TechWorld's* headline at the time announced,

---

[*] The newly constituted regulatory body set up as a result of the 2003 Communications Act.

"Ofcom to Throw Radio Spectrum Wide Open: Market Forces It's Over to You," pointing out in the article that followed that "Ofcom is planning the biggest shake-up in telecoms regulation since it began 100 years ago" (Judge, 2004).

The 'Spectrum Framework Review: Implementation Plan' (January 13, 2005) which followed the consultation process, moved swiftly to set out the timetable for transition to spectrum trading and liberalisation in relation to mobile services and also outlined the release of further radio spectrum that it would make available to the market for mobile services in the following two to three years. Spectrum trading, which allowed licences to be bought and sold without referral to Ofcom, was to be implemented almost straightaway, although some caution was applied to change of use (liberalisation), which continued to require Ofcom's approval. Once the process of deregulation had been set in motion, the momentum for greater market determination of spectrum allocation continued apace. Ofcom's 2009 consultation document, 'Simplifying Spectrum Trading: Regulatory Reform of the Spectrum Trading Process and Introduction of Spectrum Leasing,' pointed out the following:

> We introduced spectrum trading in the UK at the end of 2004 as a key element in our programme of market-based reform. Since then, we have progressively extended trading to a broader range of licences and plan to continue doing so. (Ofcom, 2009, section 1.3)

Ofcom's proposals at this stage were to amend the 2004 trading regulations to remove the need for Ofcom's consent altogether for most classes of licences and, in terms of spectrum leasing, to remove the requirement for Ofcom to be advised of details of leasing arrangements or to issue a new licence. Each of these key elements introduced a further level of detachment between Ofcom and the management of the radio spectrum, allowing market forces and interests to take control of the process with minimum regulatory involvement.

Ofcom reported a positive response from respondents to the proposed changes, suggesting there was general agreement that the current trading process was "impeding desirable spectrum market developments" (Ofcom, 2010, p. 6) and plans were announced to proceed with the introduction of these new trading rules. In February 2011, new regulations were introduced that would allow spectrum trading for the first time in the mobile bands (900MHz, 1.8GHz and 2.1GHz), which Ofcom argued would promote more efficient use of the spectrum and allow operators "with a greater need" to bid for frequencies from those who were under-using it. Finally, in March 2011, Ofcom published its plans for the long-anticipated auction of 4G (fourth generation) mobile phone spectrum, the rights to which are in demand to enable the faster connection speeds needed for smartphones and tablet computers. The auction process, which includes the 800MHz bandwidth or 'digital dividend' released as a result of the switch-over

Kathy Walker

from analogue to digital television broadcasting, is due to start at the end of 2012 with bidding commencing early in 2013.

Issues and Implications

The pressure on the radio spectrum has been growing steadily in the last decade as a result of growing demand for spectrum use from the mobile phone industry and, more recently, the mobile broadband sector. The International Telecommunications Union (ITU) data indicate that, "by the end of 2009, there were an estimated 4.6. billion mobile cellular subscriptions, corresponding to 67 per 100 hundred inhabitants globally" (ITU, 2010, p. 1). By 2011, the worldwide number of active phone accounts had grown to an estimated 5.2 billion (Consultant Value Added, 2011). In the UK alone, Ofcom estimated that by June 2011, there were 80 million mobile phones in use, 12.8 million of which were smartphones (BBC, 2011c). The speed of growth of the mobile phone industry and the pervasive nature of mobile phone use internationally has generated a "trillion dollar" business generating $1.2 trillion from sectors including handset sales, services (mostly earned by the mobile operators/carriers), network infrastructure sales and phone accessories (Consultant Value Added, 2011). Revenues from associated industries such as SMS and MMS mobile advertising almost doubled from 2008 to 2009, generating an estimated $8.8 billion in 2010 (Consultant Value Added, 2011).

It is not difficult to see, therefore, how the fastest growing industry in the world has had such an impact and influence on national governments' policies for the deregulation of radio spectrum management. Each new technological innovation, such as the current explosion of wireless technologies, promises economic revival, employment growth and the resurgence of consumer spending, all of which links spectrum policy to key government policies in these areas. In many ways, the features of the deregulatory processes outlined earlier reflect the diminishing confidence of national governments in their right to manage the radio spectrum on behalf of their citizens and the abdication of these responsibilities to the imperatives of the market. It is also indicative of the unequal levels of influence the corporate communications industries have on the policy-making process in an era of continuous technological change and the changing balance of power between these industries and national governments in shaping government priorities in relation to spectrum allocation. The process demonstrates an inherent belief in the market to deliver the best solutions to the changing needs of society. Ofcom's (2009) statement that, "spectrum trading allows spectrum to be transferred through the market to those that can generate the greatest benefits for society and so helps secure optimal use of the limited and valuable spectrum resource," suggests the market mechanism itself can, in some way, determine, and therefore generate, the greatest benefits for society, rather than those for the corporation or its shareholders. There is no role here for the measurement of the public interest, illustrating as it does a naive conflation between what might be perceived as economic and social benefits.

The influence of Cave's 2005 analysis of the economic priorities for spectrum management and regulation and the Cave and Webb (2004) research report, 'Wireless Communication Policies and Politics: a Global Perspective' are clearly evident in the framework of the consultation. This economic determinant for deregulation is also present in many other reports and consultancy documents outlining suggested spectrum 'reform' strategies. Wellenius and Neto's (2007) working paper, which looks at the reform of spectrum management in developing countries, for instance, begins its analysis with the statement that, "moving management of the radio spectrum closer to markets is long overdue"(p. 1). Although both technical and economic efficiency remain key factors to be taken into account when addressing the needs of the changing communications sector, the current preoccupation with economic incentives, driven by the growing demands of the expanding group of companies that make up the mobile telecommunications industry, should not be allowed to displace the needs of the wider social and cultural community. As the analysis report for the European Commission makes clear, "Ideally spectrum should be distributed efficiently, which means given access to the combination of uses and users that maximise economic value-added, subject to taking account of social welfare and public policy concerns" (Analysis Consulting and Partners, 2004, p. 14).

For this reason, a number of questions need to be asked about the colonisation of the policy-making process by commercial interests and the marginalisation of social welfare and public policy concerns. Perhaps the most obvious question to be asked when considering the future of a precious public resource like the radio spectrum is 'whose voice is heard' in the discussions about its future and how representative are the views submitted in response to Ofcom's consultation documents? As would be expected in relation to a major reorganisation of policy in the communications sector, and in keeping with the standards of accountability expected from a statutory regulatory body, Ofcom's consultation process is set up as an open and public process organised to generate a wide range of views and responses. Ofcom's consultations are carried out in accordance with its consultation guidelines, which are published online, and Ofcom acknowledges that, "consultation is an essential part of regulatory accountability the means by which those people and organisations affected by our decisions can judge what we do and why we do it" (Ofcom, 2007).

In relation to the 2009 consultation that preceded the policy document, 'Simplifying Spectrum Trading: Reforming the Spectrum Trading Process and Introducing Spectrum Leasing' (2010), Ofcom reported that a "broad spread of sectors and stakeholders" (p. 6) had responded to the consultation. However, in total, this amounted to just 11 responses, all industry based, including six commercial spectrum-using organisations of which three were from the programme-making and special events(PMSE) sector, two mobile network operators (MNOs), two trade associations and one consultancy. Considering the background of the respondents, it is not really surprising that, as Ofcom reported, "The responses generally agreed that the present trading process is impeding desirable spectrum market developments and supported many of the changes proposed" (Ofcom, 2010, section 1.7). Some respondents, including a communications infrastructure and media services company and a provider of radio communications system and coordination software, urged Ofcom "to go ahead faster or further" with the proposed changes (Ofcom, 2010, section 1.7). It seems that Ofcom and the industry in general are in broad agreement with the deregulatory policies suggested, but the wider voice of the public is missing.

There are several possible reasons for this lack of response from the wider community. Firstly, although Ofcom's consultations are published on its website and there are dedicated spectrum technology and policy sites, which track and publish details of worldwide spectrum developments, there has been relatively little coverage in the general media about the thrust of policy deregulation in relation to the UK radio spectrum. In addition, Ofcom publishes so many consultations in relation to the rapidly changing communications sector and its deregulatory proposals that it is difficult for those public and campaigning organisations that might have an alternative view to respond to each consultation document. The Voice of the Listener and Viewer (VLV), for instance, did respond to the *Review of Radio Spectrum Management: an Independent Review for UK Government* (2002) carried out by Cave in relation to the specific impacts of spectrum liberalisation on broadcasting (VLV, 2002). However, the silence of the media in terms of coverage of these key public issues is deafening. In general, the lack of published information and media coverage of the key changes occurring in relation to management of the radio spectrum as a whole undermine both public awareness of the issues and any possible informed debate or response.

Another key public policy concern in relation to the growing role of market mechanisms in radio spectrum allocation is the use of auctions as a means of assigning spectrum usage rights. Jeremy Bulow, former chief economist of the FCC's Bureau of Economics, has argued that, "auctions are a good way for governments to sell public goods like spectrum because it leads to the person who values the spectrum the most, the person with the best business plan, getting it" (cited in O'Toole, 2001). His argument that "auctioning generally leads to higher prices than negotiating"(O'Toole, 2001) is borne out by recent examples. When India auctioned spectrum for 3G telecom services in 2010, nine telecom companies participated in the online auction process, generating $15 billion for the Indian government (IntoMobile, 2010). The auction was so successful that the Indian government almost immediately opened bids for Broadband Wireless Access (BWA) spectrum, essential for enabling internet access for mobile devices, raising a further $5.48 billion US. The financial rewards of these auctions for cash-strapped governments are clear, but how far they ensure that the use of the spectrum auctioned meets the long-term and possibly unpredictable needs of society in the future is not so clear. In relation to the UK 3G auction in 2000, analysts have argued that the mobile operators overpaid for their licences, and as a result, did not have the funds to invest in the infrastructure (BBC, 2011b). In the long run, this slowed the development of the 3G networks and resulted in higher prices for UK consumers.

Significant concerns have been raised about the proposed 2012 4G auction, arising from the release of analogue television spectrum, and how it will be utilised to benefit society in general. Analysts have warned that the auction is likely to be complicated, making it both difficult to understand and expensive take part in, and will therefore advantage existing telecom companies over bidders who might have more social uses for the spectrum such as providing rural broadband coverage (Judge, 2010). Tony Lennon, president of the broadcasting union, BECTU, who has described the forthcoming auction as "the biggest release of a public asset in a generation," also argues that because of the government's attitude that "the most profitable solution is by definition the most socially useful" (Judge, 2010) the successful bidders would almost certainly be the telecommunications operators. Most worrying is Ofcom's decision during the consultation process to shelve its original proposal to require one of the successful bidders for this 4G spectrum to build a network capable of serving at least 95 percent of the country. This obligation has now been built into a separate government-funded scheme, but since, under present licensing arrangements, some rural parts of the UK do not even have mobile voice services, it is highly likely that there will continue to be unequal access to vital communication services in rural areas. It is debatable whether commercial operators of 4G services will match the government's proposals for expanded voice and broadband coverage if it is commercially unprofitable and there is no regulatory requirement to do so.

The process of spectrum liberalisation promulgated by Coase and Cave, and the move to market mechanisms in the allocation of radio spectrum, could be described as a success in economic terms. However, far too often in these accounts, the rationale for traditional government spectrum management is explained solely because the spectrum is a limited resource that needed to be allocated for specific uses and users. Cowhey, Aronson, and Abelson (2010) suggest that the traditional justification for government's central role in spectrum management "was that radio spectrum constituted a scarce public resource that could be degraded by radio interference among competing uses" (p. 175). However, just as the justification for public service broadcasting is often reduced to a simple technical explanation of 'spectrum scarcity,' which ignores the sociocultural impetus for its existence, the traditional role of government policy in the allocation and management of radio spectrum owes as much to a complex interaction of social, cultural and economic considerations as it does to technical limitations. A further problem resulting from the process of liberalisation therefore stems from the prioritisation of the carrier over the content provider as a result of the failure to distinguish adequately between the value of different uses of the spectrum. In the US, television and radio broadcasters who supported spectrum regulation and are still its most vocal defenders are declining in economic significance in relation to other industry sectors. As Hazlett (2001a) suggests, "according to a 1992 FCC estimate, some UHF television stations in Los Angeles were only 1/20th as valuable as cellular telephone systems using the same frequency space." This change in the balance of economic power between the telecommunications and broadcasting sector is also evident in the UK, and many critics of Ofcom's strategy for the liberalisation of the radio spectrum argue that its policy does not sufficiently recognise and acknowledge the cultural and social applications of the spectrum for broadcasting applications when compared to telecommunications applications. The Radio Authority criticised the "one size fits all" approach of the 2002 Cave review of spectrum management arguing that, "although it may be helpful to establish a valuation of spectrum, it does not follow that broadcasters should necessarily pay that valuation for their licences . . . the public interest obligations on ILR (Independent local radio) [are] an appropriate return for the use of the spectrum" (Radio Authority, 2002, p. 2).

These criticisms of the economic determinist approach to spectrum liberalisation, which stress the differences between broadcasting and other uses of the spectrum that do not have the same public service content obligations, are particularly important in relation to the release of the spectrum resulting from the analogue to digital switchover. Significantly, proposals for the sale of part of the 800MHz spectrum for mobile telephony and the coexistence of these new services with existing digital television provision are likely to interrupt the signals for an estimated 2 million terrestrial television viewers (Ofcom, 2011a; *The Telegraph*, 2012) presenting a significant threat to the principle of universal provision of public service broadcasting. The BBC has also expressed concerns about its ability to deliver HDTV services within the limited spectrum allocation available on the digital multiplexes set aside for DTT. Research carried out on behalf of the BBC Trust by Sagentia suggests that, "at a technical level, between one and four HDTV channels could realistically be provided within the existing DTT allocation through six Multiplexes" (Klein, Reynolds, & Johnston, 2007, p. iv). These six multiplexes include both the BBC and commercial providers of DTT channels, and the restrictions this technical limitation places on the free-to-air services they can offer to audiences places them at a substantial disadvantage in relation to pay television platforms such as BSkyB and Virgin Media. The report points out that most other countries including the US and leading European nations have committed to providing HDTV on their terrestrial platforms, which suggests that in their haste to free up the analogue television spectrum for alternative commercial uses, the UK government's policy in this area has significantly undermined the future development and competitive position of public service broadcasters and the interests of the audiences they serve. Even on an economic level, Sangentia's research suggests that, "given that one of the rationales for digital switchover was to retain and enhance the UK's lead in TV technology, it would be ironic to erode that lead in a transition to HD"(Klein et al., 2007, p. 25).

Conclusion

This chapter has outlined the changes that have taken place in the policy for allocation of the radio spectrum and some of the issues it raises. It has examined what appears to be a fundamental shift in attitude to the custody and management of the radio spectrum as a public good and the commercial pressures that have influenced the way it is currently conceived of as a tradeable commodity. In a headlong flight to release additional radio spectrum for the mobile broadband sectors, governments around the world have made available large tranches of radio spectrum previously utilised for broadcasting, military and civil applications. Although the process in the UK has been marked by numerous cycles of regulatory reports, consultations and "stakeholder" responses, which Ofcom argues are designed to secure long-term benefits from scarce spectrum resources, the pressures for rationalisation and redistribution of the spectrum have largely been driven by the leading mobile telecommunication operators and the anticipated additional profits to be gained from new network services and mobile data revenue. Some liberalisation of the radio spectrum is undoubtedly necessary to facilitate the development of next-generation mobile broadband, but Ofcom's light-touch regulatory approach to the practice of auctioning spectrum may undermine the benefits to society more widely than these developments could bring. Licences for the 4G spectrum will be awarded on a similar basis to previous auctions, with awards made on a UK-wide basis and for an indefinite duration, with very limited powers for the regulator to revoke a licence during the first twenty years. The licences will also be technology and service neutral (that is, determined by the operator) and permit all types of spectrum trading without the need to return to the regulator.

This market-driven, light-touch regulatory approach has inherent problems, not least the extent to which Ofcom has relinquished regulatory control of this spectrum to the commercial sector. As previously argued, it is difficult to predict future applications and demands for this public resource. The FCC noted in 1997 that, "no government agency . . . can reliably predict public demand for specific services or the future direction of new technologies" (Hazlett, 2001a) and even Cave and Webb's (2004) pro-liberalisation approach acknowledged that, "forecasting future demand for wireless is notoriously difficult because of the changing applications and technologies in this area" (p. 20). Regulators simply do not know what future applications might be or even how they will be affected by regional or international decisions on spectrum allocation, suggesting that the drive for investment and competition should not override the need for continued regulatory oversight. Ofcom's original proposal to impose minimum service requirements to ensure that both urban and rural areas benefit from mobile broadband developments appears to have been shelved in favour of a yet-to-be-determined government initiative and call in to question the extent of their commitment to the public interest when it clashes with industry imperatives. Without regulatory requirements in place, the commercial priorities of companies bidding for additional spectrum allocation are unlikely to include development and provision of services to less-accessible parts of the country.

There is also concern for the UK DTT sector, which is already disadvantaged in terms of the spectrum available for development of HDTV services and its impact on the ability of public service broadcasters to compete in the increasingly competitive and commercialised multi-channel television environment. This is exacerbated by the recent acknowledgement of the serious impact that the coexistence of 4G mobile services on the 800MHz band could have on existing free-to-air terrestrial television services. In conclusion, Ofcom's stated principle that, "it is preferable to look to market mechanisms to promote the efficient use of resources rather than regulatory intervention, unless the case for such intervention is clear" (Ofcom, 2011a, p. 6), underestimates both the value of the radio spectrum as a public good and the role of the regulator in ensuring that a wider range of social and cultural responsibilities are given equal consideration alongside market imperatives when determining its future allocation.

TEN

Secondary Schools Under Surveillance

Young People 'As' Risk in the UK. An Exploration of the Neoliberal Shift from Compassion to Repression

Charlotte Chadderton

This chapter addresses the repositioning of marginalised youth 'as' risk (Giroux, 2009), rather than 'at' risk, an arguably relatively recent phenomenon across western Europe and the US. In this chapter, I consider the increase in the use of new surveillance technologies in schools as an example of the shift from 'care to control' (Harvey, 2003), which characterises neoliberal systems of governance, to explore the possible impact of such extensive school surveillance on social inequalities. This is considered in the wider context of the creeping securitisation and militarisation of education in general, particularly in the US and, to a lesser extent but still significant, in the UK. I innovatively apply Judith Butler's thinking on 'performativity' to consider the way in which discourses of surveillance shape the subjectivities of different social groups in order to analyse how some young people are constructed as 'threatening.' I argue that although the installation of surveillance systems tends to be justified in terms of enhanced security, such school surveillance simply contributes to the creation of disadvantaged populations on the margins of society. As the study of new technologies of school surveillance is in its infancy in the UK, this chapter also highlights the need for further research on the impact of these devices on the lives and perceptions of those involved.

Youth as Risk

> . . . the destruction of the welfare state has gone hand in hand with the emergence of a prison-industrial complex and a new carceral state that regulates, controls, contains, and punishes those who are not privileged by the benefits of class, colour, immigration status and gender. (Giroux, 2009, p. ix)

Young people have historically occupied an ambiguous position in society. Whilst sometimes regarded as a threat, they "have historically been linked to the promise of a better life" (Giroux, 2009, p. ix) as well, and regarded as a positive investment for the future. However, this has been changing over the last forty years or so, leading some to argue that youth today is viewed as a danger to society, particularly disadvantaged youth. Rather than being considered as a group that is 'at' risk—vulnerable to social and economic disadvantage and exclusion—they are more and more often positioned overwhelmingly 'as' risk themselves. The changing position of youth has come about in a specific social and economic context. Over the last forty years, the global economy has been transformed, prompting talk of 'the new capitalism' (Sennett, 2005) and 'the new economy' (Castells, 2002). This process started with the oil crisis in 1973 and consequent global recession and was exacerbated after 1989 and the fall of the Soviet Union. States have removed restraints in place on the free play of market competition (Bauman, 2004, p. 51). This 'new capitalism' has been accompanied by a change in governance, often referred to as neoliberalism (Dean, 2008). The post-war welfare state, which would previously have protected disadvantaged groups from destitution, is in the process of being—at least partially—dismantled. This process of reduction in government spending on public services is referred to by Harvey (2003) as 'accumulation by dispossession' and has led to the marketisation of the 'living space' (Harvey, 2006). Public services are cut and private business is championed. Indeed, this process of transfer of wealth from public services to private coffers has sped up considerably in the UK since the bailing out of the banks in 2008–2009 with public money and the austerity measures of the coalition government that came to power in 2010. The task of national government has become "the promotion of economic efficiency and competitiveness" (Dean, 2002, p. 53), and all other activities of government are evaluated "first in terms of the availability of resources, and second as to whether they contribute to or inhibit economic efficiency" (Dean, 2002, p. 53). In the UK, this is resulting in the marketisation and privatisation of public services such as education, at the expense of any social purpose, a notion which, to many, would have been unthinkable a decade earlier (Ball, 2007).

The new economy tends to be associated with discourses of inevitable global 'progress,' which is seen as impossible, and even nonsensical, to resist. Bauman (2004) argues that a "profit-orientated . . . society of consumers" (p. 39) has been created in the West. However, the new economy has equally led to the creation of large communities of "unemployable and invalid" (Bauman, 2004, p. 51) people who would previously have worked in heavy or manufacturing industries but for whom there is now little or no work. Giroux (2009) argues that, "[p]eople are no longer citizens but consumers" (p. 2). The 'society of consumers' creates a social arrangement in which the 'unemployable' cannot fully participate, and their lack of economic participation is further positioned as a wider lack of social participation. The prevalence of such discourses of inevitable progress allows states to abdicate responsibility for their policies (Skeggs, 2004, p. 79) and the 'unemployable' are positioned as suspicious for their alleged lack of social participation (Bauman, 2004; Dean, 2002, 2008; Skeggs, 2004).

Skeggs (2004) argues that the culture of the most disadvantaged is pathologised in order to legitimate the reduction of the welfare state and blame the poor for their own disadvantage, allowing them

> to be identified as *the* blockage to future global competition and national economic prosperity. The identification of a pathological cultural hindrance to modernity is the means by which structural problems are transformed into an individualised form of cultural inadequacy. (p. 79)

Bauman (2004) argues that the new capitalism is producing similar large swathes of disadvantaged populations across the globe, for whom there is no work in the new economy. These disadvantaged populations are understood as 'failed consumers,' unable to take part in the consumer society, and as such, "failed neoliberal citizens" (Monahan & Torres, 2010, p. 4). Bauman (2004) refers to these economic victims or failed citizens as perceived 'human waste,' suggesting that the production of such 'waste' is an integral, inescapable part of modernisation:

> The production of 'human waste', . . . is an inescapable side-effect of *order-building* (each order casts some parts of the extant population as 'out of place', 'unfit' or 'undesirable') and of *economic progress* (that cannot proceed without degrading and devaluing the previously effective modes of 'making a living' and therefore cannot but deprive their practitioners of their livelihood). (Bauman, 2004, p. 5)

As the 'new capitalism' is positioned as an inevitable system to which there is no alternative, those who do not participate are seen not only as a deliberate hindrance to social inclusion, but particularly in the UK, also as deliberately hindering the progress of the nation: "It is the character of individuals and their willingness to fit into the nation that count" (Skeggs, 2004, p. 84).

Skeggs (2004) argues that over the last fifteen to twenty years, as the working class has suffered growing levels of unemployment, Britishness as an identity has been redefined as 'middle class,' and working-class identities have been reconstructed as "non-respectable who represent a threat to civilisation, citizenship and, ultimately, global capitalism" (p. 91). Skeggs suggests, then, that the working class has been redefined as beyond, or even threatening, to Britishness.

Young people in disadvantaged communities are being hit particularly hard by the conditions of the new economy, as job prospects are very poor "for the school leavers who enter fresh on a market concerned with raising profits through cutting labour costs and asset-stripping rather than creating new jobs and building new assets" (Bauman, 2004, p. 10). In July 2011, UK youth unemployment had reached almost one million (13.5 percent of 16–24-year-olds), the highest level ever recorded in the UK (Kingsley, 2011; ONS, 2011). By media and politicians alike, youth tend to be especially demonised (Osler & Starkey, 2005; O'Toole, 2007), portrayed as feral, irresponsible, incapable of learning and possibly violent and criminal. In this way, growing youth disadvantage can be blamed on the young people themselves, social and economic inequality portrayed as a problem for the individual rather than of wider social structures. Indeed, there is a new version of a very old argument in British welfare politics that claims the existence of an 'undeserving poor,' whose destitution is considered to be of their own making.

> In this context, it is argued that young people, rather than occupying their previous ambiguous position between hope and risk, tend no longer to be linked to the promise of a better life in the future but actually become positioned as a threat or risk themselves: "In a radical free-market culture, where hope is precarious and bound to commodities and a corrupt financial system, young people are no longer at risk: they are the risk." (Giroux, 2009, p. x)

Giroux (2009) argues that for these groups, whom he refers to as "disposable populations" (p. 8), the categories of 'citizen' and 'democratic representation' no longer apply, or apply very differently, as I discuss here.

Proponents of neoliberal governance typically claim they support the idea of a small state and in UK Prime Minister David Cameron's words, the "big society." However, the retreat on welfare and public services is actually coupled with an increase in security and control—thus in reality, a 'big state,' but big in security terms rather than welfare. This combination of market liberalisation and security, summarised by Gamble (1994) as the politics of "the free economy and the strong state" has been a key element of 'New Right' thinking in the US and UK since the 1970s and was exemplified by the Reagan and Thatcher governments in the 1980s. The focus of state expenditure is shifting away from care and toward control (Harvey, 2003), that is, away from meeting human need and increasingly toward the surveillance and control of suspect populations.

> Repression increases and replaces compassion. Real issues such as a tight housing market and massive unemployment in the cities—as causes of homelessness, youth loitering and drug epidemics—are overlooked in favour of policies associated with discipline, containment and control. (Giroux, 2002, cited in Bauman, 2004, p. 85)

The poor and disadvantaged are now subject to either reform or simply surveillance and containment by state mechanisms in order to improve national economic performance (Dean, 2002, p. 55). Indeed, this return to force and so-called paternalism can be considered integral to the neoliberal project (Dean, 2008, p. 35). Whilst both public and private funds are invested in this project of control, the state also makes a vital ideological investment. Discourses that pathologise the lower classes help drum up popular support for the 'discipline, containment and control' in the form of increased policing and surveillance. Young people bear the brunt of this, the policing justified by youth's positioning as 'feral' and therefore in need of containment.

It should be noted that a focus on youth as risk is not completely novel. There have previously been moral panics around youth behaviour in the UK and elsewhere, for example, around young, working-class men in gangs (Cohen, 1973) and young black men (Critcher, 1993); certain groups of youths have frequently been depicted as a threat to social order and referred to as 'delinquent.' In some ways, the recent characterisation of youth as 'feral' is a continuation of this tendency. However, what makes this characterisation a concern today is the way in which it serves the neoliberal project.

At the time of writing, the threatening behaviour and policing of youth in England is a hot topic of debate. In the summer of 2011, unrest erupted across the country after the police shot and killed Mark Duggan in Tottenham, North London. The events provide an appropriate illustration of the move from welfare society to discipline and control in the UK. Rather than understanding this unrest as an inchoate response to generations of police violence and discrimination and growing poverty in an increasingly unequal society (Reicher & Stott, 2011), both the mass media and politicians have condemned these protests and riots as "mindless" (e.g., Lammy, 2011), "criminality pure and simple" (Cameron, cited in Reicher & Stott, 2011) and those involved as a "feral underclass" (Clarke, cited in Wilson, 2011), or they have blamed them on the work of organised gangs (Arnot, 2011). Sentences for involvement in the unrest, which spiralled to incidents of looting and arson, have been extremely and unusually harsh, processed by emergency twenty-four-hour court sittings. The latest reports suggest that sentences for rioters are, on average, 18 percent longer than normal for similar offences (Lewis et al., 2011). In a landmark case, two men have been jailed for incitement to riot, for posting an invitation on the social networking site Facebook, although no riots occurred as a result of their invitation (Lewis et al., 2011). As I argue elsewhere, in this way, those involved in the unrest have been dehumanised and positioned as beyond the pale of 'civilised' society (Chadderton & Colley, 2012). Indeed, Members of Parliament have ironically stated for the press that the perpetrators cannot be seen as belonging to 'our' society; such statement is devoid of any recognition that many people do feel that they exist very much on the margins of the cities where they live, a situation currently being exacerbated by massive cuts to essential support services and chronic lack of training and jobs, especially for the young (see Khallili, 2011). This situation resonates

Charlotte Chadderton

with the notion of the 'failed neoliberal citizen' and the creation of communities of disadvantage, pushed further into deprivation and to the margins of society by harsh punishments for protests. For those who belong to the most disadvantaged communities, their positioning as beyond citizenship can be seen as a justification for the reduced rights and legal protection, whilst at the same time, their reduced rights position them outside full citizenship.

In this chapter, I consider one aspect of this growing surveillance: that of the new surveillance devices being installed in secondary schools across the country. This chapter, then, responds to Henry Giroux's call for

> commentaries about how . . . schools, and other educational sites in the culture provide the ideas, values and ideologies that legitimate the conditions that enable young people to become either commodified, criminalised, or made disposable. (2009, p. xii)

Surveillance in Secondary Schools

It is in this context of a move from care to control, of the replacement of compassion with repression, that I consider the introduction of new surveillance technologies in UK secondary schools, as an exploration of one of the ways in which this regime of securitisation is actually playing out. Although there have been periodic shifts between a state focus on compassion and repression since 1945, the move to repression has intensified over the last thirty years. Recent years have seen an insidious securitisation and militarisation of (urban) life and culture, not only in the fast-growing cities of the global south or those of the conflict-torn Middle East, but also in the US and UK (Graham, 2011). It is argued that the same technologies of security and control developed by, for example, the US and Israel, during armed conflicts abroad, are now being employed to control Western populations. One aspect of this securitisation is increased surveillance, particularly in the US and UK. There has been an explosion of various kinds of new surveillance technologies installed across UK society. The UK has more Closed Circuit Television (CCTV) cameras than the whole of Europe together—an estimated 4.2 million (Minton, 2009). Indeed, this massive increase in surveillance is a phenomenon within Europe, as yet mainly restricted to the UK.

The securitisation of societies is pervading both private and public spaces, including educational institutions (Saltman & Gabbard, 2011). Educational regimes in the US and UK are becoming increasingly severe. Student misdemeanours are punished very harshly: levels of both temporary and permanent exclusion from secondary schools remain particularly high in England, despite permanent exclusions having fallen by two fifths in the last six years (The Poverty Site, 2011). Schools are more frequently involving the police in punishments for pupils that would previously have been dealt with by staff, detentions or meetings with parents. Increasingly, on-site police officers have appeared (see Kupchik & Bracy, 2009), most common in the US, but still present at least part time in some, particularly inner-city UK schools. Some would argue that teachers need protection from increasingly violent young people in their classrooms and assistance with searching pupils for knives, drugs and alcohol (Garner, 2008), and one study conducted in Scotland (Black, Homes, Diffley, Sewel, & Chamberlain, 2010) suggests that the presence of a police officer does tend to improve staff–pupil relations and feelings of security. However, in the US, there are reports of physical abuse (Kupchik & Bracy, 2009), and pupils have been arrested for misdemeanours as small as being late, wearing the 'wrong' clothes or throwing paper airplanes (McGreal, 2012).

The increase in new technologies of surveillance, then, can be seen as an aspect of this wider securitisation of education. As in the US, schools in the UK have installed CCTV, metal detectors, alcohol and drug screening programmes, chipped identity cards and electronic registers, biometric surveillance such as iris and fingerprint recognition, cyberspace surveillance including webcams, internet logs, websites hosting student data for parental access and auditing programmes including student databases and threat assessment software, among others (Hope, 2009). There have been reports of systems that log the food a pupil selects for lunch and send the information to a website for parents to check their child's diet (UK Press Association, 2009) and even of CCTV cameras being installed in school toilets (Chadderton, 2009). Furedi (1997, cited in Hope, 2010), referring at the time to the US, although these words could now be applied to the UK, argues that, "many schools look like minimum security prisons" (p. 231). Although there has been much discussion devoted to these new technologies and their impact in general, as an educational phenomenon, new technologies of surveillance in schools is only just beginning to receive media and academic attention in the UK (see, e.g., Hope, 2009; McCahill & Finn, 2010; Taylor, 2009). More research has been conducted in the US, where the debate is more widespread than in the UK (Monahan & Torres, 2010).

The installation of surveillance devices tends mostly to be justified on the grounds of security (Hirschfield, 2009; Marx & Steeves, 2010), but reasons of health and personal welfare are also frequently given regarding surveillance in educational institutions. Protection from both external and internal threats of 'dangerous others' has been cited, along with fear of attack, including the murder of London secondary school head teacher Phillip Lawrence at his school gate (1995) and the massacre in a Dunblane Primary School (1996) and also school and college shootings in the US such as Columbine (2001) or Virginia Tech (2007). The reduction of bullying, theft, knife crime, smoking, junk food consumption and truanting are equally frequently cited. In these discourses of security, the young are positioned as 'at risk' and in need of protection.

However, evidence suggests that surveillance systems do not actually ensure security: For example, there was both an armed guard and a video surveillance system at Columbine. How else, then, can we understand such extensive surveillance? What impact do new surveillance technologies have in schools? Whilst such extensive surveillance of young people raises a myriad of questions, it is beyond the remit of this chapter to discuss them all, and the focus of this collection is inequality. Evidence from the US does suggest that for many, surveillance does calm fears and panics, and the effects of this should not be underestimated (Monahan & Torres, 2010). However, as Ragnedda (2010) argues,

> [s]urveillance is much more than simply monitoring, watching and recording individuals and their data . . . . Surveillance is an interaction of power that creates and advances relations of domination. In practice, surveillance is a mode of governance, one that controls access and opportunities. (p. 356)

Schools have always built surveillance and control practices into their organisation. These include physical observation, attendance registers, classroom arrangement, dress codes, behaviour policies, streaming, assessments and exams (e.g., Foucault, 1991). However, it is likely that these new devices leave less unobserved space, both physical and discursive, than staff and pupils have previously experienced. The new technologies should, though, not be viewed as separate from other forms of surveillance in schools; rather they should be considered "within embedded power relations" (Simmons, 2010, p. 55).

It is often presumed that surveillance is neutral, affecting all of those under surveillance equally (Monahan & Torres, 2010). However, as I have argued elsewhere (Chadderton, 2012), more recent work on surveillance in general has identified the 'social sorting' (Lyon, 2002) function of surveillance regimes. A special issue of the journal *Surveillance and Society* (2008) focussed exclusively on surveillance and inequalities, demonstrating that surveillance is experienced differently by distinct social groups (Monahan & Fisher, 2008). In the special issue, the focus is innovatively the subjects of surveillance, and the authors argue that surveillance has a norming, regulating effect and assists in the sorting of the population (Monahan & Fisher, 2008; Monahan & Torres, 2010).

There is a lack of research mapping the installation of new surveillance devices in schools in England. However, it is likely that they have been unevenly installed (Hirschfield, 2009). An Economic and Social Research Council-funded project on the 'surveilled' (McCahill & Finn, 2010), conducted at the University of Hull, examined the social impact of 'new surveillance' technologies on the lives of 13- to 16-year-old children. The researchers examined three schools and found that most CCTV cameras had been installed at the school with the poorest intake of students, and no cameras at all had been installed in the private school in the study. They also found that young people of lower socioeconomic classes were more likely to feel they were the target of surveillance, and they were more likely than middle class young people to adjust their behaviour when they felt watched, including staying away from spaces under surveillance such as shopping centres. Middle-class young people did not tend to feel they were the target of surveillance; rather they reported feeling that surveillance devices were installed for their own security. Equally, females were more uncomfortable with surveillance than males and less likely to resist the surveillance by reappropriating control and conducting the surveillance themselves, as their male colleagues did, by, for example, filming friends on mobile phones. It is, therefore, likely that not only are the devices unevenly distributed, they are also very differently perceived, with perception changing according to the social group. Although these findings come from a small study and cannot be considered generalisable, it seems they support the view that surveillance will enhance existing class and gender inequalities. They also fit in with other research on the way in which the installation of new surveillance technologies reinforces existing inequalities (e.g., Langstone, 2009).

As I argue elsewhere (Chadderton, 2012), when considering the social category of race rather than class—although empirical research linking new technologies of school surveillance and race has yet to be conducted in the UK—it is likely that school surveillance impacts more harshly on racial minorities than whites. Minority ethnic groups are already disproportionately subjected to more surveillance outside school, for example, stop and searches on the streets, airport controls and police profiling, "which continue to rely upon racial markers of 'risk.'" (Monahan & Fisher, 2008, p. 217). Minority ethnic groups are seven times more likely than their white counterparts to be stopped and searched, and stop and searches on black and Asian people have increased by 70 percent in the last five years (Travis, 2010). Young black people represent 15 percent of the total youth population in London but account for 37 percent of all stop and searches of young people, 31 percent of all of those accused of a crime by the police and 30 percent of all of those dealt with by Youth Offending Teams (Race for Justice, 2006).

Equally in a school context, minority ethnic young people are more likely to be excluded from school than white young people (Gillborn, 2006; Parsons, 2008). Particularly in the current counterterrorism context, racial minorities, particularly those who are perceived to be Muslims, are already positioned as embodying threat and in need of control and surveillance. As racial minorities are already frequently positioned as threatening, we can assume that these discourses are likely to build on longstanding notions of perceived essentialised links in the UK between minority ethnic bodies and criminality and threat (Ragnedda, 2010). Evidence from the US suggests that the presence of CCTV cameras in schools may indeed be impacting more harshly on racial minorities. Hirschfield (2009) reports an incident in which a school principal suspected drug dealing in his school through his observations with CCTV cameras. He invited the police into the school, who conducted a raid that involved holding weapons to students' heads; however, no students were arrested. Although African American students only made up 25 percent of the school's population, they constituted the majority of those involved in the raid. This raises important questions about how African American students are perceived and the impact of such incidents on the way in which they perceive themselves. I now move on to consider how new technologies of surveillance might impact upon students' subjectivities.

Subjectivities and Inequalities

The work of Foucault (1991) on the panopticon is frequently used to theorise issues of surveillance, allowing us to understand the modern, Western system of governance, which functions by causing subjects to internalise specific values and therefore self-regulate because they imagine themselves to be under constant surveillance. Subjects take on what is referred to as 'the gaze' of the powerful and discipline themselves, behaving according to a given society's norms and values and thus contributing to the normalisation of these values. Foucault refers to this process as the creation of 'docile bodies,' individuals who manage themselves. However, as Simmons (2010) argues, Foucault's concern is the way in which regulation operates, rather than the groups that are regulated, and he did not differentiate among these docile bodies.

I therefore suggest that the possible impacts of surveillance can be theorised drawing on the work of Butler (1993, 1997, 2004, 2010), in order to explore the way in which discourses around notions of surveillance and control shape the roles and identities of students and youth practitioners. Butler's work is useful for scholars exploring the way in which subjectivities are constituted, allowing a critical, in-depth study of the way in which identities are produced and reproduced through the internalisation of norms. For Butler, like Foucault, identity categories do not reflect essential or innate subjectivities. Rather, an individual is 'subjectivated,' rendered a subject, through discourse. Butler's work builds on Foucault's by arguing not only that discourse forms the subject, but that multiple discourses are at play, which position different bodies in different ways, by which I mean that expected behaviour is different for female bodies than for male bodies, for example. Discourses communicate what is normal and not normal, for people from a given class, ethnic group or gender in any given society. In this way, discourses actually regulate identities. The subject is both constituted and constrained by subjectivation.

Most importantly for understanding the impact of school surveillance, Butler argues that the constitution and reconstitution of identities functions on a day-to-day basis through a practice she calls 'performativity.' By this she means that gender, race and other identities are something we 'do,' not that we are, and we perform our identities, mostly unwittingly, in different ways in different situations—but mostly according to accepted social norms in a process of self-regulation. The word 'performativity' should not be confused with a 'performance': No identity is considered more 'real' than another. Identities are seen as inherently shifting and multiple but still very regulated by normalising discourses. Performativity is not a single, deliberate or conscious act. These repeat performances construct, confirm and perpetuate given social norms.

> Consider the medical interpellation . . . which shifts an infant from an 'it' to a 'she' or a 'he', and in that naming, the girl is 'girled', brought into the domain of language and kinship through the interpellation of gender. But that 'girling' of the girl does not end there; on the contrary, that founding interpellation is reiterated by various authorities and throughout various intervals of time to reinforce or contest this naturalised effect. The naming is at once the setting of a boundary, and also the repeated inculcation of a norm. (Butler, 1993, p. 7)

Bodies are equally viewed as produced through discourse, "in its surface and its depth, the body is a social phenomenon" (Butler, 2010, p. 33). The body (and the way it moves, dresses, speaks) defines how we are understood, in terms of gender, race, class, sexuality and (dis)ability:

> The human is understood differently depending on its race, the legibility of that race, its morphology, the legibility of that morphology, its sex, the perceptual verifiability of that sex, its ethnicity, the categorical understanding of that ethnicity. (Butler, 2004, p. 2)

Discourses need not be explicit or spoken in order to shape the frame through which identities and interaction are understood—indeed, the regulation of subjectivities is invisible. Thus it becomes clear that the discourses surrounding surveillance—who is under surveillance and for what reasons—will impact on the students and their self-perceptions and also on practitioners who work with young people.

Butler's work suggests that it is through the performing of norms that an individual internalises his or her identity and position in the social hierarchy and will mostly self-regulate by acting according to socially accepted norms. This is very relevant when considering the way in which young people, particularly those who belong to marginalised groups, have been repositioned as risk or threat to contemporary society. Indeed, US research suggests that, "disproportionate policing and surveillance of urban minority students functions to prepare such students for their rightful positions in the postindustrial order, whether as prisoners, soldiers or service sector workers" (Hirschfield, 2009, p. 40). Thus Butler's work enables us to understand better how surveillance operates to shape subjectivities and perceptions.

Those who are already positioned as at risk, such as the white, economically secure middle classes, will have this positioning reinforced by increased surveillance, and those who are positioned as risk, black, Asian, and working-class young people, will mostly be positioned as risk. As the discourses that constitute the subject pre-exist the subject, an individual subject is perceived as the embodiment of the discourse (Butler, 2010). In the discourse of youth as risk, groups of young, economically inactive people have become the (imagined, nevertheless, with real consequences) embodiment of threat. The invisibility of this process is likely to further normalise existing social inequalities.

Drawing on Judith Butler, then, I argue that the procedures for allegedly ensuring the security of young people at school in the UK are actually reproducing structures of inequality. The installation of new surveillance technologies can be seen, at least partly, as a response to discourses of marginalised youth as risk, and in addition, feed into and reproduce these discourses. Surveillance procedures actually reproduce privileged bodies as privileged and marginalised bodies as threat.

The positioning of young people as threat is dangerous in many ways. In January 2012, 15-year-old Texas student Jaime Gonzalez was shot dead by police at school for brandishing a pellet gun. Officers believed the gun was real, but the standoff lasted only thirty minutes and no attempts were made to simply injure the boy to debilitate him (Robbins, 2012). The incident raises important questions about the readiness of the police to shoot to kill a child in school, the performing of threatening behaviour and the embodiment of a young person as the ultimate threat. The presence of on-site police officers (although not present at Jaime Gonzalez's school) means young people are criminalised for behaviours that would not have been so had a police officer not been on-site (Kupchik & Bracy, 2009). Whilst on-site, police officers can, in fact, have an inclusionary function (Hirschfield, 2009); it has been argued that minority ethnic and low-income youths are disproportionately affected by police presence (Kupchik & Bracy, 2009). Extensive surveillance renders misdemeanours more visible, and young people are more likely to be caught.

Indeed, it is likely that such extensive surveillance contributes to what is referred to in the US as the 'school-to-prison-pipeline': the very close link for many, mostly males of colour, between school and incarceration. In the US, it is estimated that 1 in 100 adults is incarcerated, and 1 in 9 African American males between the ages of 20 and 34 (Monahan & Torres, 2010). It is equally well-documented that at present, the UK locks up more young people than any other EU country (BBC, 2009a). The age of criminal responsibility stands currently at 10 years old, with only Switzerland (seven) and Scotland (eight) lower in Europe (BBC, 2010a), and the number of children held in prison on remand has risen by 41 percent since 2000–2001 (Pemberton, 2010). Again, this raises questions about the embodiment of young people as threat, even criminal, the framing of their difficulties as criminal problems and the extent of systems put into place for them to be removed to the margins of society.

Of course, there is the possibility of resistance to such surveillance. The notion of subjectivation developed by Butler allows the subject certain agency, and indeed, evidence suggests that some young people are resisting this extensive surveillance. In 2009, pupils at an East London school walked out in protest after CCTV cameras were installed in their classrooms without consultation, claiming their civil liberties had been infringed (Colasanti, 2009). Hope (2010) describes the ways in which children still find spaces where they are unobserved, clicking away inappropriate websites when their teachers come close and using other students' passwords to log in, in disguise. Hope also reports how students hack into computer systems to access information about staff. However, the majority of Hope's examples involve male students, frequently viewing or distributing pornography and thus performing dominant forms of masculinity. Equally, in McCahill and Finn's (2010) study, those more likely to resist were males, the resistance involving filming other students themselves. This seems to imply that even in much resistance to surveillance, gendered identities are reaffirmed and inequalities perpetuated. Although Butler has been criticised for underestimating agency, these cases do suggest the only possible agency is "a radically conditioned form of agency" (Butler, 1997, p. 14).

It seems likely, then, that in the context of the neoliberal agenda in schools, new technologies of surveillance regulate and control bodies accordingly, and that the existing race, gender and social class dynamic in schools is reinforced by increased surveillance, feeding into a system that encourages the young to uncritically accept their position in the social hierarchy and practitioners to treat them accordingly. There is a need, therefore, to look beyond Foucault (1991) and the panopticon (McCahill & Finn, 2010) in order to understand the impacts of surveillance. Surveillance will be experienced as bound up with the performativity of identities and the ways in which an individual experiences his or her own body and the ways that body is positioned by others. Identities are already gendered, raced and classed, and these positions are reaffirmed by surveillance. "Surveillance technologies, therefore, possess degrees of agency such that they do not simply uncover pre-existing truths but actively contribute to the creation of certain truth regimes" (Monahan & Fisher, 2008, p. 218).

The Implications for Democracy and Citizenship

In the wider socioeconomic context, where public welfare services are being cut, there is a widespread attack on civil liberties (Giroux, 2009) and refusal to accept dissent; public wealth is moving to private ownership, public spaces are privatised and militarised and it is clear that democratic systems are under threat. New technologies of surveillance in schools largely seem to play directly into this new regime, functioning, at least in part, as a way of policing marginalised youth for whom there are few opportunities in the new global economy, rather than supporting or investing in them. "Surveillance often operates as a mechanism for the management and exclusion of individuals within neoliberal regimes" (Monahan & Fisher, 2008, p. 218).

Research has argued that those who are perceived to be in need of surveillance are positioned as suspects (Douglas, 2009; Monahan & Torres, 2010). If marginalised young people are no longer regarded as (future) citizens, but as suspects, this equally has implications for the democratic system (Giroux, 2009). In this way, such increased surveillance begins to change the relationship between state and citizens as their rights are reduced. Indeed, this takes on extra importance when we consider that young people as a group lack political representation and are, as such, legally disenfranchised. A recent study conducted in the north of England (Taylor, 2009) argues that such surveillance in schools equates to a loss of the right to privacy for young people.

As mentioned earlier, neoliberalism in the UK has involved the redefinition of Britishness as middle class, thus implicitly excluding disadvantaged communities from inclusive understandings of citizenship. This is reified by the extensive surveillance and policing these communities experience: They are less entitled to the protection of law than other, more privileged groups. In the current context, marginalised young people's lives become increasingly precarious (Giroux, 2009).

> . . . the circuitry of social control redefines the meaning of youth, subjecting particularly those marginalised by class and colour to a number of indiscriminate, cruel and potentially illegal practices by the criminal justice system. In the age of credit and quick profits, human life is reduced to just another commodity to be bought and sold, and the logic of short-term investments undercuts long-term investments in public welfare, young people, and a democratic future. (Giroux, 2009, p. x)

Are we seeing the beginning of the creation of large groups of 'human waste' in the UK, communities beyond citizenship, made up of disadvantaged, disenfranchised and over-policed young people?

Conclusion

Given that public education putatively supports the progressive goal of equality,

> the use of surveillance to target and sort students along lines of race, class, and gender deserves continued scrutiny and critique, especially as the institution of education further aligns itself with the criminal justice system, the military, and private industry. (Monahan & Torres, 2010, p. 2)

In this chapter, I have argued that the widespread installation of new technologies of school surveillance is both fed by, and reinforces and reproduces, social inequality. The act of surveillance produces inequality as well as is produced by inequality. Viewed in the context of neoliberal governance and the new capitalism and the wider securitisation of society, extensive school surveillance can be seen as an integral part of the shift from care to control in the UK, justified by popular discourses of individual deviance, deficiency and threat. I have argued that youth, particularly youth marginalised along lines of class, race and gender, are positioned increasingly as a threat to national progress, which further justifies their containment and policing—this is clearly illustrated by the politicians' response to youth unrest across England in summer 2011. Although security no doubt plays an important role in schools' decisions to install surveillance systems, such extensive surveillance cannot be considered without reference to wider sociopolitical developments. Indeed, the security of the white, middle classes is an important part of the neoliberal agenda and will serve to further marginalise those who do not belong to these groups. The use of a Butlerian framework highlights the way in which young people from marginalised groups are understood as the embodiment of the threat.

This chapter also serves to illustrate how further research needs to be done in this field, particularly on the securitisation of education in the UK, including research that explores the perspectives of the students themselves, as well as those of the practitioners and the local communities, and the question of how the data collected by the surveillance devices are used.

ELEVEN

Reclaiming the Media

Technology, Tactics and Subversion

Maxine Newlands

Technological advances in media production have created a convergence of media platforms that are apparently open to all. Technology enables us to consume and communicate across the globe, and this has brought certain benefits for capital. This globalized new technology also brings political conflict closer to home, however. For example, journalists credit social media with helping to organise the uprisings of the Arab Spring, and this new technology has brought human rights abuses in Syria and Bahrain and protests in Tehran to the attention of diasporas and traditional news media in the West. This is remarkable, particularly given that as Morozov (2011) notes, social media is less a "Twitterocracy "(p. 247) and more a tool that by itself does not necessarily "herald an era of transparency and honesty" (p. 136). In the UK, however, social media has also brought benefits and problems for protest movements.

This chapter explores how new social media enables protests to bypass mainstream media practices and facilitates the production of media by activists themselves. It examines how the internet echoes the structure of protest movements and shows how a simple website can be a powerful organising tool. New media enables activists to invert standard media practices and utilise everyday media practice as a tactic in influencing public opinion. The chapter concludes by looking at what problems are raised with this innovation in relation to the digital divide and an over-reliance on social media by activist groups. It draws illustrative examples from the environmental activist movement in the UK in order to highlight the positive and negative elements of technology. The environmental activist movement in the UK is the focus of this chapter and provides a useful case study in that activist groups have been early adopters of the internet and through their use of new media technologies, they have realised the potential to move local protest onto the global stage.

First, we consider new media technologies and how social movements use technology to reinvent earlier protest strategies and tactics. The following section looks at how websites and Web 2.0 have specifically enabled activists to invert media practice and to produce their own media. The chapter concludes by looking at what cost comes with embracing new media technologies.

New Media Technologies and New Social Movements

The internet was initially created as a communication tool by the American military; it has subsequently aided the advance of both capital accumulation and global communication. Creating new forms of communication and developing new technologies as part of "capital's dream of superfast networks that will spread consumerism across the planet" (Notes from Nowhere, 2003, p. 65), the internet and later the World Wide Web (in the 1990s), gave new opportunities for wider communications, greater networks and organisational structures—not just for capitalists but also for protest movements. The internet changed modes of communication, from one-to-one (such as the telephone), one-to-many through print and broadcasting media, to many-to-many (Gilmore, 2006).

The relationship between the web and mainstream media practice led to an increased interaction between media houses and the general public. Consumers could now react quickly to a story or news event. More co-operations emerged, and there was less suspicion between the general public and the media. The web also shifts the way news is consumed. With the development of smartphones, tablets and the internet, news can be consumed 24 hours a day, from anywhere in the world. No longer do consumers have to wait for the six o'clock headlines or next day's newspaper to gain information. The result is that the internet meant journalism became an "old practice in a new context—a synthesis of tradition and innovation" (Kawamoto, 2003, p. 4). Technologies enabled consumers to access news at any time. Moreover, the internet makes journalists out of everyone. As well as consuming the news, the web and the internet enable citizen journalism (i.e., through blogs, video apps and smartphones) but often at the cost of diminishing the authority of traditional journalism in ways that are not always desirable (Newlands, 2010).

In addition, it must be acknowledged that as Gillmor (2006) notes, "the development of the personal computer may have empowered the individual, but there were distinct limits" (p. 16). Such limits are defined by the concept of 'digital divide.' Digital divide is a term that emerged in the mid-1990s and refers to the inequality between those who have 'ever,' and those who have 'never,' had access to the internet. This inequality can be polarised through educational opportunity, wealth, age, urban/rural location and physical ability. The internet also brings with it divides in democracy. Although journalism may have been opened up, it was still regulated and moderated by the organisations and individuals that had access to their websites. However, those who have heavily utilised the internet for civic engagement are the individuals involved in the new social movements associated with environmental activism.

The growth of the internet has worked to the benefit of new social movements, by expanding their numbers and aiding the easier coordination of tactics and skills. The year 1999 saw the explosion of technologically-led social movements across the globe. Following protests outside the World Trade Organisation meeting (on December 1), the world awoke to media images of protesters and rioters clashing on the streets of Seattle, Washington. Newspaper images of handcuffed activists, tear gas clouds and police standing over protesters were already familiar in the UK. Six months earlier, a similar event took place in London, with the Global Justice Movement's J18 protest, Carnival of Capital. Seattle and London were united by what appeared to be a spontaneous anti-capitalist protest that emerged from nowhere. In reality, the two events were a highly organised protest as a "result of clear sets of mathematical principles and processes that govern a highly connected network" (Notes from Nowhere, 2003, p. 68). The protest had been coordinated through a network of internet sites and e-mail messages.

Whilst the creation of the internet was a facilitating device for consumer capitalism, it also meant that activists could flourish in the "public part of cyberspace" (Lovink, 2002, p. 254). This growth in communication meant activists' networks in the UK could learn from other activists' movements around the world, as "global networks of power and counter-power landed simultaneously to confront each other in the spotlight of the media" (Castells, 2009, p. 340). The most notable example is the Zapatista movement, which partly inspired the Global Justice movement. Briefly, the Zapatista movement emerged when Canadian, Mexican and American governments drew up the North American Free Trade Agreement (NAFTA), framing it as an opportunity to lower trade barriers. However, it led to lower subsidies to the indigenous populations, whilst opening up opportunities for large corporations. To appease their indigenous population, the Mexican government agreed to an amendment to NAFTA, and when newly elected President Fox sent the Indigenous Rights Bill to be passed in 2001, the Zapatista army (a group of farmers) travelled the 2,000 miles to the capital to address congress. When the army reached Mexico City, it was greeted by 100,000 people. Bringing thousands of people together through the internet showed the potential organising possibilities of the web. By 2001, details of the Zapatista's protest against NAFTA spread around the world via e-mail, websites and blogs.

These movements were effective because "broad-based, local and national networks, run by communities, and linked internationally, by the internet, have proved themselves capable of bringing together very large groups of people in very short spaces of time" (Kingsnorth, 2003, p. 75). Networks of activist movements develop over the internet, because there is a symbiotic relationship in the organisational structures and networks of both the internet and established protest movements.

New social movements and users of cyberspace involve interactions among like-minded people, with shared interests, often operating in non-hierarchical ways. Indeed, as Jenny Pickerill (2003) notes, "cyberspace has been likened to that of a rhizome" (p. 24), in that a "rhizomatic structure provides multiple entryways, facilitating potential participants' entry into environmental activism through connections to their rhizomatic online networks" (p. 24). The hypertextual architecture (Kahn & Kellner, 2003) of the internet has been referred to as a non-hierarchical 'rhizome' (Deleuze & Guattari, 1988, p. 7). This is a linear network that connects any point to another point, understood in terms of a non-signifying system that is neither singled down to one aspect or multiple aspects (Deleuze & Guattari, 1988). The rhizome has been defined as

> [A] multiplicity that has no coherent and bounded whole, no beginning or end, only middle from where it expands and overspills. Any point of the rhizome is connected to any other. It has no fixed points to anchor thought, only lines; magnitudes, dimensions, plateaus, and they are always in motion. (Deleuze & Guattari, 1988, p. 377)

This rhizomatic approach helps us understand why the internet might be an appropriate communication tool for protest movements, as it is not hierarchically structured or organised, mirroring the makeup of radical protest movements. The term 'radical protest movements' refers to protest collectives that exist outside of NGOs, single issue protest movements or lobbyists, such as the vertical network of Greenpeace. Vertical networks, which favour linear developments, are not often found in new social movement organisations and environmental activist collectives. Rather, protest movements are often characterised by horizontality. Actions are arranged through consensus politics, which is why the term 'rhizome' is useful in understanding how technology is used by activist movements. E-mails, blogs and forums are all used in the decision-making process, alongside face-to-face meetings. So, the internet echoes the rhizomatic networks that shape the environmental protest movements and makes it an attractive technology for protest. In addition, new technological developments in modes of communication narrow the division between mainstream and alternative media forms. The effect is that activists have a new platform to bypass the traditional media with its often unsympathetic messages and produce their own websites, blogs and media.

It also means they are able to produce their own media on a global scale—important examples here would include http://www.indymedia.org.uk or http://www.schnews.org.uk.

Furthermore, by enabling activists to bypass the traditional media, activist media is free of the 'order of discourse' that favours states over activists, which means that activism and protest stories are often unfavourable to protest movements. The creation of alternative media (such as Indy media) means that activists can be both producers and consumers of news. The symbiotic relationship between activists and the internet shifts any action from local event to, potentially, global news. Websites mean activists can provide information directly to journalists, document protests themselves and post their own coverage. As Castells (2009) notes, besides the more positive messages about protest generated by Indy media, numerous hacklabs, temporary or stable, populated the movement and used the superior technological savvy of the new generation to build an advantage in the communication battle against their elders in the mainstream media (p. 344).

This has led to new techniques that draw on the notion of 'cyberlibertarianism' and the ability to develop "electronically mediated forms of living with radical libertarian ideas about the proper definition of freedom, economics and community" (Heath & Potter, 2004). Indeed, many use the internet for both physical and virtual activism. Virtual and viral activism consists of such actions as 'cyber squatting,' where hundreds of people log onto a corporate website and crash it in protest of the company's activities; 'hacktivism,' which is the fusion of politics and activism and involves the creation of open-source software; 'blogging,' being the democratic expression of networking, also a media critique and journalistic socio-political intervention; and 'Google bombing,' which is redirecting users to subversive sites. Activists are reworking the technology by placing hackers and cyber squatters at the "forefront of the movement, freeing activism from the limitations imposed on their autonomous expression by corporate control of the media networks" (Castells, 2009, p. 345).

Thus activists have utilised the rhizomatic, horizontal architecture of the internet, new technologies, smartphones and Web 2.0, in order to organise their protests through virtual and physical networks. The internet has become a pivotal tool for organising protests, informing the media or voicing opinion, and it has become a "key ingredient of the environmental movement in the global network society" (Castells, 2009, p. 316). The web provides the tools to enable activist movements to develop their own media and political strategies, and it has extraordinarily "improved the campaigning ability of environmental groups and increased international collaboration" (Castells, 2009, p. 316). Activists are now able to use a new "global communications infrastructure for something completely different, to become more autonomous" (Notes from Nowhere, 2003, p.

65). The capacity of new technologies to support and sustain dispersed coalitions of protestors and new forms of political organisation has been witnessed in the anti-capitalism protests (J18 in Seattle in 1999; the May Day protests between 2000 and 2004) and similar 'summit sieges' at the G8 Conference in 2005 and G20 Conference in 2009.

Smartphones, BlackBerrys, and 3G and 4G mobile telephone technology also make use of "computer networks for purposes different from those assigned by their corporate owners" (Castells, 2009, p. 345). For example, the 2009 Camp for Climate Action used an earlier activist's technique of the 'swarm' but organised it via SMS, e-mail and blog spots, to coordinate a 'swoop' on a pre-designated organisation in central London that was deemed to have a poor environmental record. Swoop and swarm are terms coined by the RAND (2002) Corporation, which began in 1946 as a research project (Project Rand) backed by the US Armed forces and undertakes research commissioned by government agencies and private firms. A swarm, like the collective actions of swallows and similar birds, is used as an analogy to the protest movement. Like swallows, many protesters move en masse: Each "moves as one, as if it's one organism. Yet no-one is in charge, it seems to happen as if magically" (Notes from Nowhere, 2003, p. 67). The RAND Corporation believes that "swarming would be the main form of conflict in the future . . . . There is enormous power and intelligence in the swarm" (Notes from Nowhere, 2003, p. 66). Previously, a swarm would have been organised through word of mouth, limited direction and vague instructions based on Chinese whispers and a maze of symbols and whistles to indicate the start of the swarm. For example, during the Carnival Against Capital (on June 18, 1990), 8,000 face masks were handed out to activists. The masks were of different colours and, on a signal (in this case, a whistle), each colour (red, blue, green or black) would follow one person with a correspondingly coloured flag out of the railway station (Tyler, 2003). Later, with the internet, this was easily organised via text and SMS. The same principle was applied to the 2009 Camp for Climate Action's swoop, but the use of new technology meant it had a greater effect.

Technology combined with a rhizomatic internet makes the organising of a swarm or swoop much easier, as SMS allows activists to arrange events through tweets and smartphone messaging. A good example of this would be the Camp for Climate Action's protest against climate change, organised at Blackheath in south London in 2009. Prior to the establishment of the camp, interested parties signed up to an e-mail list and were requested to give their mobile phone numbers. Any interested parties (including the police and journalists) were sent a series of texts that directed individuals to gather at pre-determined locations around London. Activists were then told to await further instructions for the location of the camp itself, via text messaging and the social networking site Twitter. The use of mobile phones and a social networking site opened up the event, enabling anyone with an interest to be part of the swoop, either physically or at a virtual level. Not only interested individuals utilised Twitter. Tweets were not just coming from camp participants—news organisations, local politicians, local residents and the Metropolitan Police were all using Twitter as a communication tool. *The Guardian* and *The Times* newspapers ran with live commentaries and blogs as events unfolded. The Sky News website had an overall view of different tweets on its front page, similar to the 'tweet deck' principle (one website that correlates all social media sites into one).

As technological developments expanded, the use of Web 2.0 meant activists could use social networking sites, text messaging, forums and blogs to communicate with wider society and the media. Thus, the rhizomatic pattern of the internet was and is useful for movements as it echoes the horizontal and networked politics of environmental activism. Smartphone and SMS mean activists can take the old tactics of protest, established in the protest camps and actions such as demonstrations used in the 1970s and 1980s and early 1990s, and transfer them to new technologies, making the organising of protests and actions much cheaper, simpler and more effective. The thrust of the chapter thus far has strongly suggested, therefore, that the internet has opened up greater opportunities for activists to communicate and organise protests. The rhizomatic pattern of the internet allows a movement to grow and expand without compromising the grassroots democratic processes of new social movements.

New Social Movements—Old Skills for New

To provide new platforms, social movements make the most of the technological developments discussed earlier. These platforms enable activists to circumvent the stalwart media practices and institutions of gatekeeping, press offices, spokespersons and so on. When it comes to reporting environmental activism, UK activist collectives such as Plane Stupid, Climate Rush, Camp for Climate Action, Dissent network and many others are actively feeding the media with stories and action via websites. Websites have turned the consumer into a producer, by giving mainstream journalists direct acccess to activists. Technology enables environmental activists to challenge existing hierarchical media houses. New media means activists can organise a protest as an image event and give reporters direct access to the 'action' through their website, texts, social networks and citizen journalism. Technology also enables activists to invert media practice. The greater the concentration of journalists and politicians, the higher the concentration of media coverage (Couldry, 2000, p. 156). One place that has a high level of media presence is Westminster, Central London. Protests held outside the Houses of Parliament or on Downing Street are close to the media hub of the Parliamentary Press Centre, home to 300 journalists, and Milbank Tower, with BBC and ITV studios. For example, the anti-war protest against the invasion of Iraq in 2003, the Countryside Alliance march in 2002, the Pensioners Rights protest in 2006 and the Make Poverty History campaign in 2005 all occurred at "sites in the media eye" (Couldry, 2000, p. 156). Activist and documentary filmmaker Hamish Campbell at the Visionon.tv website, for example, encourages activists to produce their own news media. Anti-aviation expansion collective Plane Stupid has a communication strategy that uses websites to provide journalists with direct access to protesters. Doyle (2009) observes how Plane Stupid's website "constitutes its action . . . the website is action orientated . . . alongside press releases" (p. 113). For example, during the protest on the roof of the Palace of Westminster (the UK parliament building), at the centre of the 'media eye' in the UK, Plane Stupid provided the mobile phone numbers of the activists involved, on its homepage, enabling journalists covering the event to talk to the protesters whilst they were protesting. A second benefit of websites is the ability to get news footage out quicker than mainstream media. As Plane Stupid activist Dan Glass notes, social media enables his organisation to bypass the state and police prevention measures:

> . . . just being able to upload something we've occupied, Manchester, Stanstead, Aberdeen airports, it's been genius, just . . . . You know, one of the things about the Aberdeen protest, they put a Fire Engine in front of the cage that we were in so that no media could see us, but by that time we'd taken our own photos and sent it out so 'I've done it anyway'. So it's been really useful. And of course there was the Twitter on the whole swoop and everything like that, so it's been brilliant in many ways for organising actions and bypassing, A), the police and B), the powers that be, and C), the traditional media. (personal communication, August 9, 2011)

Such use of new media also provides a platform for citizen journalism. In a mediatised world, activists can produce their own news stories by adopting mainstream media practices in the virtual world (via online press releases, hypertext links and media websites). This blurs the boundaries between the traditional media and new media and between journalists and activists. Knowledge is exchanged instantly between activists and journalists. In discussing a protest against a proposed third runway at London's Heathrow Airport, Plane Stupid Activist John Stewart argues that this enables activists to shape the way in which protest is represented:

> . . . I think certainly potentially, it [social media] helps us to set the agenda, in a way we didn't before, on the roof of the House of Commons, the Plane Stupid was taking the pictures. Those pictures were going to Sky, BBC before anybody else could, so, so Plane Stupid set the agenda, their version of things and I think it helped in the immediacy of organising. (personal communication, August 9, 2011)

Activists realise that although the internet helps build global activist networks, there is also a necessity for using space both physically and symbolically. Stewart further notes

> . . . when the five or six people who went onto the roof of The House of Commons, they were very clear in the early meetings that we didn't want headlines like that [points to London Evening Standard language of militant to describe activists] so even to the extent of what they would wear. A decision was taken that they would dress up like the sons and daughters of Daily Telegraph readers to attract those papers . . . . It so happened that the Daily Mail actually came to them and wanted to do a feature. (personal communication, August 9, 2011)

The intention was to invert the media eye that focuses on Parliament, to challenge the status quo. As Stewart puts it, "[we] try and use the media by subverting it, by using their own tools" (personal communication, August 9, 2011). Plane Stupid deliberately attempted to subvert the media, by taking the protest to the media eye to gain coverage from the many journalists and permanent television cameras around Parliament Square. In one sense, such action could be a heterotopia. Although the action gave Plane Stupid front page coverage, much of the language focused on a discourse of terrorism. The now defunct London newspapers—London Lite and London Paper—both ran with an image of the protesters on the roof. Headlines called the protest a "Storming of Parliament" (Murphy, 2008) and "The Graduate Eco-Warrior in Commons Raid" (Mendick, Murphy, & Low, 2008). The London Paper announced "Airport Protesters Make a Mockery of Commons Security" and offered this headline: "Security Alert at Commons: Protesters Scale Parliament" (Sutherland, 2008).

This coverage highlights another issue, that technology and new media struggle to challenge existing systems and discourses. Despite giving direct access to the protesters via the website, however, the coverage was in less than favourable terms. Although environmental activists can "increase their chances of enacting social and political change—even if they start from a subordinate position in institutional power, financial resources, or symbolic legitimacy" (Castells, 2009, p. 302), there needs to be a note of caution. As Hamish Campbell notes, whilst social media "are the best tool we have ever had":

> I think activists relying on them wholesale with such . . . naivety that I want to scream and jump . . . and the success of the corporate stuff comes at the price of the failure of the radical alternative stuff and the radical, alternative media is in the doldrums nowadays. I mean, Indy Media almost doesn't exist. (personal communication, August 9, 2011)

Morozov (2011) argues the following:

The problem with political activism facilitated by social networking sites is that much of it happens for reasons that have nothing to do with one's commitment to ideas and politics in general, but rather to impress one's friends. (p. 186)

The internet shouldn't be a solitary force for mobilisation of new political forms, but one factor, and "success is conditioned by many factors that have little to do with the internet" (Gavin, 2009, p. 130). Activists also need to be aware of the digital divide and the need to think long term for any movement to survive. John Stewart comments as follows:

> . . . back to the days when we didn't have social media, certainly not Twitter, even before fax, I've got to admit we were still able to organise. We had to plan much further in advance. We possibly had to sit and think through our strategies more than we are forced to right now because we know we can rely on some immediate media now and it's not as if we don't think through our strategies. I think we had to think fairly long-term and we had to plan in advance, so I think we had no choice but to do that . . . I mean the other concern I've still got about some of the new media is that everybody is not yet on it. There's a divide and it's not just, you know, an older generation but it's also, I think, people who have access to education and to money. And if you don't have access to the money and to the education to be able to use and read the new media, then there is a danger that you are not part of it, and we as campaigners and activists can be as guilty as anybody else of leaving those people behind. (personal communication, August 9, 2011)

In a time-conscious environment, the media rely heavily on public relations sources, and this could be extended to relying on protest websites:

> The pressure placed on them [journalists] to produce additional web-based copy alongside conventional packages—with fewer resources and an infinite amount of time—can lead to a dependence on readily available PR sources that, some argue, compromises the quality and integrity of the resultant coverage. (Gavin, 2009, p. 136)

Castells (2009) argues that "internet networks are essential to bring together the hundreds of local organisations and the thousands of activists come to the local from the global" (p. 325). Activist Dan Glass echoes this sentiment:

> I don't think it's the question that social media is the problem in terms of inter-cultural organising for political change. I just think in Britain, we don't have that cultural, intercultural organising; people stay in their different issues. Whereas in America you can say there is a lot more overlap between racial justice, environmental justice, gender justice, de, de, de, de, you *use* social media for the context whereas here, we're all fighting our own battles and not. And now I think there is a change, and this is what So We Stand is about as well, it's joining the dots and seeing the power structures and I think social media could be used for that,

for intercultural organising, I can't really see why not. (personal communication, August 9, 2011)

Therefore, whilst organising over the internet can be of benefit for global connections among social movement organisations, it needs to be combined with "face-to-face interaction" (Castells, 2009, p. 342). Activists have come to realise that in order to influence the front page, they need to "concentrate less on online communities and realise that we are communities ourselves . . . do more face-to-face stuff and, whatever means necessary, we are going to go to the actual mainstream media" (Steve, activist focus group, personal communication, August 9, 2011). Morozov (2011) observes that:

> While Facebook-based mobilization will occasionally lead to genuine social and political change, this is mostly accidental, a statistical certainty rather than a genuine achievement. With millions of groups, at least one or two of them are poised to take off. But since it's impossible to predict which causes will work and which ones won't, Western policymakers and donors who seek to support or even prioritize Facebook-based activism are placing a wild bet. (p. 180)

There is an element of digital divide that highlights the limitations of social media. Moreover, there is an element of "speaking to that audience and speaking, preaching to the converted" (Nim, activist, personal communication, August 9, 2011). It can be used to subvert the traditional media but is too limiting in terms of demographics—it reaches a limited audience. In aiming for the front pages through new media, activists can widen the demographic reach of their message but not necessarily engage some of the more socially and economically disadvantaged within activist ranks.

Conclusion

New social movements and new media offer new opportunities to bring together political activists and journalists. The emergence of technology such as smartphones, Web 2.0, the internet and blogs provide a new platform to bypass mainstream media practices. New media also allow for a protest to potentially shift from a locally covered issue to a global phenomenon. Twitter and Facebook can generate political interest. Equally, however, such social media sites can give an illusion of a greater strength or support than might exist in reality.

The range of protest organisations in the UK associated with the environmental movement have been one of the best examples of social movements that have embraced social media. Social media provides a platform to speak directly to journalists and the public and to bypass the traditionally conservative and statist politics of traditional television and print news media. Technology enables new social movements to reinvent and adapt old tactics of physical protest such as demonstration and occupation. Yet, there are difficulties in that these new media tactics are most likely to engage those from relatively privileged social backgrounds and lead, potentially, to a 'digital divide' amongst communities of potential protesters. A wider repertoire of protest strategies shouldn't be forgotten in favour of technologically-led, quick-fix solutions to 'getting the message across.'

TWELVE

Conclusions

New Horizons and Contested Futures

Erika Cudworth, Peter Senker and Kathy Walker

In this book, we have attempted to consider some societal, economic and technological developments in a broad historical, political, social and ideological perspective. The world capitalist system has been primarily responsible for rapid economic growth over the last two hundred and fifty years, and this has been closely related to the rapid technological change that the system has stimulated. The resulting enormous increase in production of an ever-changing and increasing range of products and services has lifted hundreds of millions of people out of poverty and deprivation; and in many respects, it has offered improved lifestyles to a large and growing minority of the world's expanding population. Schumpeter (1954) argued as follows:

> The capitalist engine is first and last an engine of mass production which unavoidably means also production for the masses . . . . It is the cheap cloth, the cheap cotton and rayon fabric, boots, motorcars, and so on that are the typical achievements of capitalist production . . . . The capitalist achievement does not typically consist in providing more silk stockings for queens but in bringing them within the reach of factory girls in return for steadily decreasing amounts of effort . . . the capitalist process . . . by virtue of its mechanism, progressively raises the standard of life of the masses. (pp. 67–68)

Schumpeter (1954) also emphasised that the engine of economic growth is innovation in products and production processes, rather than price competition:

> . . . markets reward mainly the large organisations which own and control massive agglomerations of land, capital, and resources devoted to innovation and product and service design. (p. 81)

Since Schumpeter wrote, the long tradition of 'worship' of economic growth has become even more dominant. The central tasks of national governments are often perceived as the promotion of economic efficiency and competitiveness designed to secure economic growth. Of course, economic growth is still extremely important for the welfare of the majority of people of the world—in particular, it is very important to the people living in developing countries and of crucial importance to those who live in the very poorest countries (Collier, 2008).

However, the worldwide focus on securing economic growth has been far from completely successful in terms of its contribution to the welfare of the majority of the world's population, amongst whom poverty and deprivation are still prevalent. Moreover, as a consequence of economic growth's enormous extravagance in terms of the use of resources and its neglect of environmental considerations, such improvements in human welfare as have been secured are at significant cost in terms of degradation of the environment of the small planet that we inhabit.

Everywhere in the world, inequality is also of great significance to the population's welfare. When analysts consider inequality, they often only consider differences in income among average living standards in various countries. This can be misleading. For example, while average living standards in rapidly developing countries such as India and China have increased very rapidly in recent years, there are still hundreds of millions of people in such countries living in poverty and suffering from deprivation. Indeed, it has been suggested that the current phase of capitalist development, which focuses almost exclusively on economic growth, is producing large swathes of disadvantaged populations across the globe, for which there is no work in the new economy.

In the West, a new profit-orientated society of consumers is being created and is generally regarded as a constituent in the inevitable progress of humankind. But the advent of this new economy has also resulted in the creation of large communities of 'unemployable and invalid' people who would previously have worked in heavy or manufacturing industries but for whom there is now little or no work. People are no longer citizens but consumers. The 'society of consumers' creates a social arrangement in which the unemployed cannot participate fully, and their lack of economic participation is liable to involve a wider lack of social participation. As has been noted elsewhere, "the more choices the rich seem to have, the less bearable to all is a life without choosing" (Bauman, 2000, p. 88). Whilst this state of affairs has been, as we have suggested, centuries in the making, the current era in the richer regions of the globe has been witness to thirty years of neoliberal restructuring, together with attempts to introduce more 'choice' via market mechanisms and practices into various forms of social provision (such as healthcare and education).

Asbjørn Wahl (2011, pp. 43–55) has argued that the 'truisms' of capitalist economics are currently rearticulated and strengthened in discourses of 'globalization' and in policies adopted by both nation states and international organisations (such as the European Union) that actively promote neoliberal forms and practices. Even successful entrepreneur George Soros (2008) has suggested that increasingly globalized practices of neoliberalism are "a greater threat to open society than any totalitarian ideology" (p. xxii). Wahl argues that this uncritical capitalist resurgence must be fought, its fundamentalisms (the very ones with which we began discussion in this book!) must be challenged and alternative futures must be seen to be possible. He quotes the radical Swedish social democrat Ingmar Lindberg who has argued the following:

> The present phase of globalised capitalism is characterized by a shift of power and a predominantly neoliberal conceptual model. It is not necessary to meet any of this with passive submissiveness. On the contrary, to be incensed at injustice is precisely the mobilizing force needed to turn developments in the opposite direction. (pp. 210–11)

Throughout this book, we have raised many questions about the current social, political and economic trajectory of capitalist development, questions that are destabilizing, at least intellectually, for the current fundamentalisms of neoliberal capitalism. This provides the rationale for our hope that it will make some contribution to answering Lindberg's call. The focus of this book, however, is on one important facet of such potentially disastrous outcomes of contemporary neoliberal capitalist development: that is, the system's extensive failures in terms of directing technological change in appropriate directions. Just as it is often assumed that the new profit-orientated society of consumers is inevitable, it is also assumed that technological change is virtually autonomous and not subject to human, social or political control. As one of our previous books, *The Myths of Technology: Innovation and Inequality* (Burnett, Senker, & Walker, 2009), demonstrated, such assumptions are largely fallacious. The failure to direct technological change appropriately is a human failure—primarily the responsibility of those who direct societies' political and economic activities.

In this book, we have tried to identify the nature of some important failures to direct technology appropriately and to outline some of the historical processes that have resulted in these failures. Most such failures can be classified under the broad general heading of 'inequalities.' Whilst an analysis of contemporary developments in neoliberal capitalism underpins all these chapters, inequalities of class, status and wealth are not the only formations of inequality with which we have been concerned. Inequalities between rich and poor regions and countries of the globe feature strongly, and these are considered in terms of (post)colonial relations. Inequalities of age, of geographic location and of ethnicity also make their presence felt. We have explored these inequalities in several contexts. In addition, we considered several aspects of an important group of technologies on which a huge amount of attention has been lavished by those—including ourselves—concerned with the analysis of relationships between society and technology. These are the information and communications technologies (ICTs). These technologies do not contribute principally or directly to meeting basic subsistence needs, but they do provide a vast new range of products and services and also offer ever more powerful potentials for increasing productivity. They also contribute to the fulfillment of needs for information, communications and entertainment. The huge attention that has been lavished on these technologies is a consequence of many factors, including sheer fascination with the activities they have made possible and the extraordinary opportunities that their exploitation has offered for making profits. This is well known, so we have concentrated on the 'hidden underside" of some of these activities.

The opportunities that computer technology seemed to offer for human satisfaction through work rather than consumption have remained largely unrealised. Some idealistic pioneers of personal computing believed that access to computers should be unlimited and total, that all information should be freely available and that computers can change people's lives for the better. They wanted everyone to have access to the wonderful tool they thought the computer to be. Their original vision was one of giving computing skills to everyone in order to enable them to reconfigure the machine to their own purpose, rather than removing the need for technical skill from the use of computers. The developers of the PC wanted the production of personal computing to be part and parcel of its use and did not conceive that technical skill would be a barrier to the democratisation of computing. Rather, they believed in cooperation and sharing possessions and knowledge, not competition, secrecy and surveillance. They opposed the increasing division of labour that would lead to de-skilling processes of production and in a concept of freedom that had been relegated entirely to the realm of leisure. That same spirit has subsequently periodically resurfaced in the world of computing, through such movements as the early stages of the development of the World Wide Web and, later, the open source software movement. But the growth of markets requires mass production. This leads to an increasing division of labour in production, the de-skilling of workers and/or their replacement by machinery to increase productivity. However, mass production also reduces prices. This democratises consumption insofar as the artefact becomes accessible to people of more modest means and with lower levels of technical knowledge. But mass production also leads inexorably to the oligopolization and monopolization of production.

The ICT industry relies heavily on rare minerals that have to be extracted from the earth. The supply chains involved in the production of commodities such as ICT devices are not apparent to the consumers of these items, and the price that manufacturers of electronic products pay for precious metals such as gold do not reflect the huge human costs paid by the miners—men, women and children—who work in terrible and hazardous conditions for low pay to mine those metals. As the gold on the surface of the earth has been mined, so men, women and children have to go deeper into the earth to extract it. Much of this mining is in countries with loose health and safety regulations—a quarter of the gold is mined by poor migrant workers in small-scale mines. The majority of accidents go unrecorded because of the itinerant nature of the workforce and the often illegal areas in which the individuals work. Buyers of ICT devices and services do not perceive the poor pay of the miners who extract these minerals, or the terrible, dangerous conditions in which they work, or the impact that mining has on these workers' health and on the general environment. Some ICT device makers try to 'ethically source' their precious metals—they buy from certified suppliers. However, there can be between four and eight levels in the supply chain of materials before those materials reach manufacturers. Manufacturers depend on the adherence of all levels in this supply chain to make claims that they buy ethically. But supply chains can never guarantee that the materials are extracted ethically. Much commercial activity now relies on networks, software and electronic transfer of capital. But this activity depends for its existence on mineral extraction and manufacturing processes previously associated with heavy industry and the repetitive work in arduous conditions that this involves.

In light of the threat of terrorism and ever-increasing demands for security, closed circuit television (CCTV) monitoring, screening technologies, tracking devices and biometric tools are used ever more extensively. New technologies of surveillance often appear to be deployed as means of policing marginalized sections of society for whom there are few opportunities in the new global economy, in preference to supporting or investing in them. But security is by no means the only issue. Implications of these developments for citizenship, social justice and participatory democracy are important contexts in which recent rapid expansion of new surveillance technologies in the community need to be seen.

The contested relationship between technologies and 'development' has been a key theme of this book. Organisations such as the World Bank try to persuade developing countries that they should invest heavily in ICTs as an engine of growth. Such propositions rely on the belief that ICTs can help developing countries narrow the gaps in productivity and output that separate them from industrialised countries—and even that they can 'leapfrog' stages of development into the information economy. But most developing countries are unable to cope with the new technological paradigm or to exploit its potential. They lack appropriate infrastructures and the capacity to adopt, adapt and innovate in their own environment. A few countries such as South Korea and Singapore have used the production of consumer electronics to transform the standard of living of a large proportion of their populations. But elsewhere, there is little evidence that ICTs can help those with no ownership and control of this technology and who do not possess the capability to innovate to achieve their development goals. Indeed, many ICT initiatives and projects in developing countries have failed to meet such objectives. Moreover, developing country consumers are influenced to deploy fashionable technologies without considering critical issues such as user capabilities and cultural feasibility. It is generally unrealistic to expect developing countries to achieve development through ICTs without policies and support in place to ensure equitable access and training, something that merely opening up markets does not necessarily ensure. In general, consuming ICTs reinforces existing social and economic inequalities.

ICT producers have promoted their products as means through which excellent education can be made available to diverse and widespread communities. The World Trade Organisation believes that public services should be opened up to international capital and that this would benefit both globalization and education. However, further opening up the markets of poor nations to transnational corporations is liable to create greater inequalities between rich and poor nations. Education is increasingly treated as a tradable commodity as opposed to a public good in wealthier parts of the globe, with learners increasingly referred to as customers, clients or consumers. In the UK, the introduction of computers into all school classrooms as part of face-to-face teaching has been promoted as the way to educate children to be the knowledge workers of the future. Yet despite the widespread integration of ICTs at all levels of schooling and into all subjects, there are still large numbers of children leaving school with low levels of literacy and numeracy. Moreover, insufficient attention has been paid to the purposes for which communities need knowledge and what knowledge is appropriate for particular groups. The use of new technologies is by no means the right answer in all circumstances. It is essential to consider what education and training are needed for the population to achieve sustainable development before considering which technologies should be used to deliver them. Governments and policy-makers need to look beyond the rhetoric of elearning and consider the best methods for educating their citizens, taking into account the local infrastructure and which modes of delivery would be most appropriate in any given situation.

The US has had a strong influence on health care policies in developing countries, just as it has in relation to development and education. This influence has been transmitted and amplified through the US's strong influence on the policies of international organisations such as the World Bank, the International Monetary Fund and the World Trade Organisation. This influence has been exerted in favour of low priority for expenditure on health care in relation to expenditure on promoting economic growth based on stimulating international trade. Such policies have tended to restrain developing countries' expenditures on the provision of clean water, which is absolutely fundamental to improving health. In developed regions of the world, resources such as clean water and refrigeration, which are also essential for conventional storage and delivery of some vital health supplies such as drugs, vaccines and injections, are readily available. In many areas of developing countries, clean water, refrigeration and other resources that are essential for safeguarding human health are scarce. The main causes of illness and death in developed countries are cancer and diseases of respiratory, cardiovascular and nervous systems. In the developing world, communicable diseases are the main health problem, and many fatal diseases could be prevented by public health measures, especially through the provision of clean water. Although some progress has been made, there is still an inbuilt tendency arising from market forces that ensures that pharmaceutical development is international and continues to concentrate on treating the diseases of the relatively affluent. Policies in developed countries, notably the US, fail to contribute sufficiently to the promotion of technological innovation in relation to the diseases that affect developing countries disproportionately. Numerous technologies are available that could offer better, more effective treatment for diseases prevalent in developing countries, but insufficient resources are devoted to development and exploitation of such technologies. US expenditure on health care per capita is immense, but this has resulted in a population whose average levels of health are poor relative to several other countries in which health expenditure is far less, but where the distribution of both income and health care has been much less unequal. Our analysis leads to the conclusion that comprehensive health care provision should have greater priority and that it is best undertaken by public authorities rather than by private organisations.

Of course, human health and welfare are of paramount importance to human beings. Animal health and welfare are worthy goals in their own right. Just as important, threats to animal health and welfare resulting from intensive methods applied to animal agriculture can threaten both human welfare and indeed the welfare of the whole planet. After the Second World War, there was a drive to produce cheap meat and animal-based food products as a means of reducing malnutrition and hunger in the US and Europe. In the 1960s and 1970s the United Nations emphasised the necessity of increasing production and availability of this food in poor countries. These initiatives were driven by the corporate interests of the multinational corporations based in the US and in Europe. They ignored the fact that pulses and grains are the most common sources of protein across the globe and that the ability of developing countries to feed their own populations successfully was significantly compromised by the replacement of staples such as corn, millet and rice by monocultures to supply the livestock feed industry. Nevertheless, the dysfunctional international system of animal agriculture seems set to expand. Western-intensive models, promoted by the agribusiness giants, are set to transform farming in some of the poorest countries in the world, just as they have transformed much of Central and South America in the latter twentieth century. The increased deployment of Western-intensive agriculture and the spread of Western food practices also have significant detrimental implications for the environment in terms of undermining bio-diversity, localised pollution, soil damage and rainforest depletion. In relation specifically to human health and welfare, inequality is very harmful to those people whom it deprives of appropriate food, one of the most basic necessities for sustaining a healthy life. Prominent in this group of about a billion people currently in developing countries are those whom poverty and landlessness deprive of the basic necessities for pursuing a healthy life—those who lack sufficient food and clean water to eliminate hunger, starvation and disease. The world's climatic conditions, soils and water availabilities are extremely diverse. It would surely be reasonable to expect that technological change would be developed, adapted and implemented so as to extract the most nutritious food from nature in light of these diverse conditions. In practice, however, the prevailing vision is of a standardized worldwide 'modern' agriculture extending from the Green Revolution to the current Gene Revolution as a standard, preferred pathway to development. Key elements of the modern world's agri-food 'system' involve a wide array of external expensive inputs such as research and development, fertilizers, seeds and irrigation. Centralized technology-driven economic growth through sustained innovation and trade is envisaged as providing pathways out of agriculture or a shift of subsistence-oriented 'old' agriculture to a modern, commercial, 'new' form of agriculture, with wider poverty reduction aims achieved through trickle-down and employment benefits from improved agriculture-led growth. This vision perceives the role of agriculture as an 'engine of economic growth' and looks forward to the economic and social transformation of the agrarian economy, from backward to modern, from subsistence to market orientated, from the 'old' to 'new' agriculture.

Alternative visions of food sovereignty emphasise working locally with natural systems, generating improved livelihoods with local agricultural production attuned to local ecological conditions. As understood by La Via Campesina, "food sovereignty requires agrarian reform in favour of small producers and the landless, the reorganisation of global food trade to prioritize local markets and self-sufficiency; much greater controls over corporations in the global food chain; and the democratisation of international financial institutions" (Branford, 2011, p. 28). Such alternatives perceive the empowerment of small farmers locally as central to achieving both economic and ecological sustainability. If the emphasis on world agricultural production were to shift to the preservation and technological development of low input systems, world food needs might be met more fully, to the great advantage of billions of small farmers and with far less environmental damage.

Extensive development in wireless technologies and applications has resulted in a growing demand for deregulation and liberalisation of the radio spectrum to allow wider and easier access for new businesses and commercial initiatives. Governments internationally have been lobbied by corporate interests to release greater tranches of radio spectrum for private development. Each new technological innovation, such as the current explosion of wireless technologies, is seen by governments as offering economic revival, employment growth and the resurgence of consumer spending. Such considerations are very influential on the formation of government policies. National governments are moving toward deregulation, reflecting their diminishing confidence in their right and duty to manage the radio spectrum on behalf of their citizens, and in doing so they are abdicating these responsibilities to the imperatives of the market. The influence of corporate communications industries on policy-making is demonstrating an endemic belief in the market as able to deliver the best solutions to the changing needs of society. These developments raise clear concerns not only in relation to the inequalities of access to the decision-making process but also for the impact this has on the priorities adopted for development of these key communication technologies. The lack of wider participation in the decision-making process and the construction of the erstwhile citizen as a mere consumer in the communications environment suggest that the likely outcome will maximise commercial applications at the expense of social goals such as universal access.

The world economic system is extraordinarily complicated, and we emphasise that it has only been possible to examine a few pertinent examples in a short book such as this: a comprehensive analysis of the vast range of world problems in relation to technology and inequality would require several expansive volumes. We also fully appreciate that the issues we consider are inevitably highly controversial. In each of the issue areas tackled by these chapters, we have tried to evaluate the evidence impartially, but we are conscious that underpinning all the chapters is a political commitment to greater equality and to revealing the way in which inequalities are embedded in everyday practices and objects. We are collectively concerned that many contemporary developments, packaged in a language of development, well-being and improvement, or choice and freedom, often do not and cannot deliver these ends and may in fact enhance existing forms and levels of inequality. Rather than making any claims for impartiality, we wish to offer this book as a serious contribution to important debates.

Bearing these qualifications in mind, the chapters in this book lead us to the conclusion that the primary purpose of the current worldwide system of production and distribution is not to satisfy the most pressing human needs but to create a never-ending stream of marketable commodities and the ever-increasing profits they generate for the owners of capital. Free competition among small-scale suppliers operates on a relatively small scale and, therefore, only benefits consumers to a limited extent. In contrast, markets reward mainly the large organisations that own and control massive agglomerations of capital that they use to exert monopolistic control over markets and exert power over governments and international organisations to structure and regulate markets in their favour. Major corporations based in advanced countries typically enjoy worldwide oligopoly and profit disproportionately from trade and technology transfer between advanced and developing countries.

In contrast, the World Trade Organisation believes that public services such as education should be opened up to international capital and that this would benefit both globalization and education. We suggest, however, that further opening up the markets of poor nations to transnational corporations is liable to create greater inequalities between rich and poor nations. These inequalities do not only relate to wealth and cannot be understood through the lens of class relations. Rather, inequalities are complex and multifaceted, involving the distribution of and access to a variety of resources (such as knowledge and nutritious food). They need to be considered in terms of relations of other social differences, gender, ethnicity and nation (e.g., in postcolonial contexts). Choices between private and public delivery systems are crucial determinants of the effectiveness of delivery of products and services to meet the needs of the whole population. We demonstrate that in spheres such as health care, education and the communications sector. The US and international organisations have exerted strong influences in favour of private delivery, and this has generally been against the interests of poor people and especially of those living in developing countries.

It is undeniable that capitalism has been responsible for extremely rapid technological change over the past two hundred and fifty years. In this and our previous books, we have presented substantial evidence that the pace and direction of innovation and technological change are fundamentally under human control and are not autonomous. Moreover, it is inevitable that any changes—whether technological or not—bring with them some benefits and some costs. Some of those costs are labeled by economists as 'externalities'—for example, costs and damage inflicted on the environment. It is beyond human powers to eliminate all the costs that technological change brings in its wake. But human intervention is capable of mitigating and reducing some such costs. It is also capable of modifying and adjusting the allocation of costs and benefits among the various groups of people involved. We contend that we have demonstrated conclusively that many such allocations of costs and benefits are generally unfair to billions of people who suffer poverty and deprivation of numerous kinds.

This book started by considering the development and exploitation of personal computers (PCs). Some of the pioneers of this now pervasive technology had idealistic dreams of the benefits this technology could bring to many people. The extent to which those dreams have been realised is minimal. But a fundamental reason for this failure may have been that those dreams were unrealisable —perhaps utopian.

Next we considered the mining of gold, just one amongst many rare minerals that form essential ingredients of PCs and numerous other sorts of ICT devices. Millions of people draw valuable benefits from the use of those devices. But the pay received by the men, women and children who mine gold and other rare minerals is very low, and the conditions in which those people work are generally appalling. There is a need for determined measures to be taken to increase these individuals' pay and to ameliorate the conditions they work in. This immediately raises the very difficult question: who is going to take the lead in taking such initiatives. We believe that within the book, there are clear general indications about the only people who can take the leading roles in driving such changes: these are the people who are most adversely affected. But in the case of gold miners, the people affected are poor and disadvantaged in numerous ways.

Nevertheless, elsewhere in the book, we find that there are some indications of long-term solutions to this problem in other sectors of the world economy. Many millions of peasants and small farmers often work for small rewards in bad, sometimes appalling conditions in agriculture. Hundreds of millions of peasants and small farmers are organising themselves in, for example, La Via Campesina, to secure some amelioration in their opportunities to be more self-sufficient by producing a higher proportion of the foods that they and their families eat. We believe that it is conceivable that, when once such efforts achieve a measure of success, other grievously exploited workers such as gold miners might be encouraged to follow their example and organise themselves to press for the changes they need to make their lives bearable. The rewards received by peasants and small farmers are generally so poor that many millions of them have migrated to cities in the hope of finding much more remunerative work in manufacturing and other occupations. To a considerable extent, such migration may well be inevitable, but we think that some improvements of the pay and conditions experienced by peasants, small farmers and their families could be secured through feasible changes in the direction of technological change and agricultural production towards "Food Sovereignty" and away from the use of intensive methods of food production, and this might secure some significant reductions in the rate of migration of people from the countryside to cities. Thinking about mining and agriculture gives rise to the idea that, in the long run, the imposition of minimum standards of working conditions and remuneration, to be imposed on a sector-by-sector basis worldwide, may be worth serious consideration. Clearly, the quantity of thought, research and lobbying needed to put such a proposition on agendas would be very substantial. In particular, research on technology and inequality in manufacturing would provide essential background to the formulation and implementation of such a proposition.

We have also shown that there are good reasons for people to aspire for radical changes in the direction of technological change in several other sectors. For example, multinational corporations are pushing developing countries to rely more on ICTs for development of their economies and for improving education, when adapting development and education to specific national needs may well be better approaches, especially for contributing to the welfare of poor and disadvantaged people. In relation to health, despite some noticeable improvements in this respect in recent years, technological change still continues to benefit rich people more than poor and disadvantaged people in most countries.

In relation to most developed countries, it has been suggested by Wilkinson and Pickett (2009) and Jackson (2011) that economic growth should no longer be an overwhelming priority. Reductions in inequality would be far more valuable for the great majority of people living in these countries. But after the 2008 global financial crisis, there was practically universal consensus on the need to get consumption and the world economy growing again. There were also calls for a global Green New Deal. Public sector investment should be low carbon, targeted toward energy security and environmental protection. Investment in such areas as retrofitting buildings, mass transit, wind and solar power and low-carbon vehicles could be designed to generate more jobs than stimuli to consumer spending. The payback from such investments would arise in the form of fuel savings and also savings in public expenditure in such areas as health costs needed to combat the adverse effects of pollution.

But the assumption behind such measures was that they would also stimulate consumption growth. Jackson (2011) suggests that, "Governments have systematically promoted materialistic individualism and encouraged the pursuit of consumer novelty . . . on the assumption that this form of consumerism serves economic growth, protects jobs and maintains stability" (p. 167). This, however, is unsatisfactory. The stimulation of consumption growth requires people to have aspirations to possess goods and services that they do not have at present, many of which they have no real need for. Growth of production and consumption in developed countries is reflected in material hungry growth in the average GDP per head. This has been based mainly on product differentiation and market segmentation and sustained internationally by enormous expenditures on advertising, marketing and promotion. Jackson maintains that investment is needed to achieve transition from a fossil fuel economy to a sustainable low-carbon economy based on renewable energy. However, he recognises that it is impossible for people to *choose* sustainable lifestyles. Such radical changes would require fundamental changes to the economy and to society.

Massive advertising and promotion, especially by major corporations and now augmented by extensive use of social media, serve to increase demand substantially. This expansion of demand helps to sustain and stimulate the drive for further increases in the GDP in rich countries. The resulting continual increased sale of products to people who have no real need for them—especially sales of luxury products—results in ever-increasing unnecessary damage to the environment, resulting both from the processes of producing the products and also from disposing of them after use. In these circumstances, it is not surprising, in the context of powerfully entrenched consumer culture, as an eminent economist suggests, that "the current environmental policies of many countries are expensive, inefficient, inhumane and in many cases entirely useless" (Sinn, 2012, p. 187). It is clearly important to further research the effects of massive advertising and promotion further as a first step in reducing them. Moreover, comprehensive analysis of issues relating to technology and inequality would include at least consideration of manufacturing, transport, energy, the environment and also social care—an important area that is becoming even more important with the growing proportion of old people in the populations of many countries.

At several points throughout this book, we have raised serious questions about the policies of international organisations such as the World Bank, the International Monetary Fund and the World Trade Organisation. It appears that those organisations' policies are often directed at meeting the needs of capitalism more than they are directed at meeting the needs of poor people in developing countries. This would seem to be an area in which far more research is needed.

It is also important to take account of arguments for the survival and continuation of twenty-first century capitalism unchanged. These are that markets offer the best incentives for individuals, companies and corporations to maximise their contribution to human welfare. Nevertheless, it is argued throughout this book that although there may has been some truth in this perspective in the past, there is very little truth remaining in it in the twenty-first century. Indeed, we go so far as to suggest that the incentives offered by markets to individuals, companies and corporations are often perverse in relation to the behaviours that are needed to secure the principal objectives of economic activities.

In light of the differing perspectives of the authors of this book, we could not conceivably arrive at a common position in relation to the precise nature of desirable futures. We all agree, however, that the future is, and must be, contested ground, and that there must be different visions and models about 'progress' and 'development.' Nevertheless, the book does constitute an argument for radical but peaceful transformations in the determination of the ways in which resources and rewards for economic activity are allocated, for significant changes in the constitutions of large enterprises and in the legal frameworks and regulations that control their activities. Above all, there is an urgent need for growing decentralization of the control of the direction of technology from huge monopolistic corporations to smaller enterprises, whether publicly or privately controlled. This need is general but perhaps most obvious in relation to agriculture and food, in which there is urgent need for far more local, small-scale production and distribution to match better with the huge biological, geographical and climatic diversity of the earth's surface. If the future is contested, it is also contestable. Whilst we may not agree on the precise trajectory of change, we are agreed that there are different pathways of development or 'progress' that might be chosen. Whilst the institutions and practices of neoliberal capitalism and its imperatives of profit seeking and increased consumption seem increasingly entrenched in our everyday lives and of global reach, they are not inevitable. The conclusions of many of these chapters may seem rather gloomy, but in fact, there is hope in each and every one, and carried by this book as a whole, that another future—of socially and culturally appropriate technologies harnessed for increasing equality and social justice—is indeed possible.

References

Adadevoh, D. (2009). *18 women, 2 men killed in Ghana gold mine disaster*. Retrieved from http://www.usafricaonline.com/2009/11/13/brknews-18-women-2-men-killed-in-ghana-gold-mine-disaster/

Adams, J., & Whitehead, P. (1997). *The dynasty: The Nehru-Gandhi story*. London, UK: Penguin.

Akrich, M. (1992). The description of technical objects. In W. E. Bijker & J. Law (Eds.), *Shaping technology: Studies in sociotechnical change* (pp. 205–224). Cambridge, MA: The MIT Press.

Alexander, S. (2001, February). *E-Learning developments and experiences*. Paper presented at the Technological Demands on Women in Higher Education: Bridging the Digital Divide conference, Cape Town, South Africa. Retrieved from http://web.uct.ac.za/org/fawesa/confpaps/alex.pdf

Alpert, M. (1990, February 26). The ultimate computer factory. *Fortune*. Retrieved from http://money.cnn.com/magazines/fortune/fortune_archive/1990/02/26/73121/index.htm

Altieri, M. A. (2005, March). Some ethical questions: The myths of agricultural biotechnology. Paper presented at the National Catholic Rural Life conference. Retrieved from University of California, Berkeley website: http://nature.berkeley.edu/~miguel-alt/the_myths.html

Analysis Consulting and Partners. (2004). *Study on conditions and options in introducing secondary trading of Radio Spectrum in the European community—Final report for the European Commission*. Retrieved from http://ec.europa.eu/information_society/policy/ecomm/radio_spectrum/_document_storage/studies/secondary_trading/secontrad_final.pdf

AngloGold Ashanti, *2010 Annual Report*. Retrieved from http://www.anglogold.com/subwebs/InformationForInvestors/Reports10/default.htm

Arestis, P., & Sawyer, M. (2005). Neoliberalism and the third way. In A. Saad-Filho & D. Johnston (Eds.), *Neoliberalism* (pp. 177–183). London, UK: Pluto.

Arnbak, J. C. (1997). Managing the Radio Spectrum in the new environment. In W. H. Melody (Ed.), *Telecom reform: Principles, policies and regulatory practices* (pp. 139–147). Copenhagen: Technical University of Denmark.

Arnot, C. (2011, November). Academic seeks new understanding of rioters. *The Guardian, 29*. Retrieved from http://www.guardian.co.uk/education/2011/nov/28/gang-culture-riots-research

Arocena, R., & Senker, P. (2003). Technology, inequality, and underdevelopment: The case of Latin America. *Science, Technology and Human Values, 28*(1), 15–33.

Arora, P. (2010). *Dot com mantra: Social computing in the Central Himalayas.* Farnham, UK: Ashgate Publishing.

Avgerou, C. (2002). *Information systems and global diversity.* Oxford, UK: Oxford University Press.

Avgerou, C., & Madon, S. (2005). Information society and the digital divide problem in developing countries. *LSE Research Online.* Retrieved from http://eprints.lse.ac.uk/2576/1/Information_society_and_the_digital_divide_problem_in_developing_countries_%28LSERO%29.pdf

Avgerou, C., & Walsham, G. (2000). *Information technology in context: Studies from the perspective of developing countries.* Aldershot, UK: Ashgate Publishing.

Ayele, S., & Wield, D. (2005). Science and technology capacity building and partnership in African agriculture: Perspectives on Mali and Egypt. *Journal of International Development, 17*(5), 631.

Baark, E., & Heeks, R. (1999). Donor-funded information technology transfer projects. *Information Technology for Development, 8*(4),185–197.

Ball, S. J. (1998). Big policies/small world: An introduction to international perspectives in education policy. *Comparative Education, 34*(2), 119–130.

Ball, S. J. (2004, June 17). *Education for sale! The commodification of everything?* Paper presented at the King's Annual Education Lecture, University of London, UK. Retrieved from http://nepc.colorado.edu/files/CERU-0410-253-OWI.pdf

Ball, S. J. (2007). *Education plc: Private sector participation in public sector education.* London, UK: Routledge.

Ball, S. J. (2008). *The education debate.* Bristol, UK: Polity Press.

Barnett, C. (1986). *The audit of war: The illusion & reality of Britain as a great nation.* London, UK: Macmillan.

Barrick Gold Corporation. (2010). *Annual report.* Toronto, Ontario, Canada: Barrick.

Bate, R. (2008). Local pharmaceutical production in developing countries: How economic protectionism undermines access to quality medicines. *International Policy Network.* Retrieved from http://www.policynetwork.net/health/publication/local-pharmaceutical-production-developing-countries

Batina, R. G., & Ihori, T. (2005). *Public goods: Theories and evidence.* Berlin, Germany: Springer-Verlag.

Bauman, Z. (2000) *Liquid Modernity.* Cambridge: Polity.

Bauman, Z. (2004). *Wasted lives. Modernity and its outcasts.* Cambridge, UK: Polity Press.

BBC. (2008, October 1). Ofcom may rethink Spectrum sale. *BBC News.* Retrieved from http://news.bbc.co.uk/1/hi/technology/7646348.stm#spectrum

BBC. (2009a, August 13). 'Too many' young offenders jailed. *BBC News.* Retrieved from http://news.bbc.co.uk/1/hi/uk/8198496.stm

BBC. (2009b, August 13). Calls to raise age of criminal responsibility rejected. *BBC News.* Retrieved from http://news.bbc.co.uk/1/hi/uk/8565619.stm

BBC. (2010a, May 22). Your money: Distance learning courses. *BBC News.* Retrieved from http://www.bbc.co.uk/news/10142490

BBC. (2010b, August 25). Chile's trapped miners told rescue could take months. *BBC News.* Retrieved from http://www.bbc.co.uk/news/world-latin-america-11092343

BBC. (2010c, March 11). Businessman jailed for organic and free-range egg scam. *BBC News.* Retrieved from http://news.bbc.co.uk/1/hi/england/hereford/worcs/8562434.stm

BBC. (2011a, March 22).Ofcom launches next-generation 4G consultation. *BBC News.* Retrieved from http://www.bbc.co.uk/news/business-12811122?print=true

BBC. (2011b, June 21). Mobile firms can trade spectrum. *BBC News.* Retrieved from http://www.bbc.co.uk/news/uk-13855655

BBC. (2011c, July 28). South African gold miners begin strike over pay. Retrieved from http://www.bbc.co.uk/news/world-africa-14324063

Bevan, A. (1952). *In place of fear.* London, UK: Heinemann.

Binmore, K., & Klemperer, P. (2002). The biggest auction ever: The sale of the British 3G Telecom licences. *The Economic Journal, 112*(478), C74–C96. Retrieved from http://www.paulklemperer.org/

Bjorkman, C. (2005). Feminist research and computer science: Starting a dialogue. *Journal of Information, Communication and Ethics in Society, 3*(4), 179–188.

Black, C., Homes, A., Diffley, M., Sewel, K., & Chamberlain, V. (2010). *Evaluation of campus police officers in Scottish schools.* Edinburgh, Scotland: Ipsos MORI.

Blair, A. (1996). *New Britain: My vision of a young country.* London, UK: Fourth Estate.

Blair, A. (2005). Keynote speech to the Labour Party Conference in Brighton. Retrieved from http://news.bbc.co.uk/1/hi/uk_politics/4287370.stm

Blanchflower, D. (2009, December 10). The cost of falling ill in America. *The New Statesman, 138*(4979), 19.

Boyle, G. (2002). Which role for ICTs in international development? Re-thinking approaches to institutional networking in Vietnam. *Journal of International Development, 14*(1), 1–11.

Braa, J., Monteiro, E., & Sahay, S. (2004). Networks of action: Sustainable health information systems across developing countries. *MIS Quarterly, 28*(3), 337–362.

Branford, S. (2011). *Food sovereignty: Reclaiming the global food system.* London, UK: War on Want.

Brett, E. A. (2000). *Developing theory, universal values and competing paradigms: Capitalist trajectories and social conflict* (Working Paper Series No. 00–02). London, UK: Development Studies Institute, London School of Economics.

Bretton Woods Project. (2008). *Agribusiness vs. food security. The food crisis and the IFIs.* Retrieved from http://brettonwoodsproject.org/art.shtml?x=561820

Brewer, M., Muriel, A., Phillips, D., & Sibieta, L. (2009). *Poverty and inequality in the UK.* London, UK: Institute for Fiscal Studies.

Brooks, A. (2006). Apple at 30—1976 to 1986. Retrieved from http://news.worldofapple.com/archives/2006/03/30/apple-at-30-1976-to-1986/

Burnett, J., Senker, P., & Walker, K. (Eds.). (2009). *The myths of technology: Innovation and inequality.* New York, NY: Peter Lang.

Burt, J. (2006). Conflicts around slaughter in modernity. In The Animal Studies Group (Ed.), *Killing animals* (pp. 120–144). Urbana: University of Illinois Press.

Butler, J. (1993). *Bodies that matter: On the discursive limits of 'sex.'* London, UK: Routledge.

Butler, J. (1997). *The psychic life of power: Theories in subjection.* Stanford, CA: Stanford University Press.

Butler, J. (2004). *Undoing gender.* New York, NY: Routledge.

Butler, J. (2010). *Frames of war. When is life grievable?* London, UK: Verso.

Cairncross, F. (1997). *The death of distance.* London, UK: Orion Business Books.

Campbell, C. (1995). The sociology of consumption. In D. Miller (Ed.), *Approaching consumption* (pp. 95–124). London, UK: Routledge.

Carlisle, S., Hanlon, P., & Hannah, M. (2007). Status, taste and distinction in consumer culture: Acknowledging the symbolic dimensions of inequality. *Public Health, 122*(6), 631–637.

Castells, M. (2002). *The internet galaxy, reflections on the internet, business and society.* Oxford, UK: Oxford University Press.

Castells, M. (2009). *Communication power.* Oxford, UK: Oxford University Press.

Cave, M. (2002). *Review of Radio Spectrum management: An independent review.* London: UK Government, Department of Trade and Industry and HM Treasury.

Cave, M. (2005). *Independent audit of Spectrum holding.* London: UK Government, Department of Trade and Industry and HM Treasury.

Cave, M., & Webb, W. (2004, October 8–9). *Wireless communication policies and politics: A global perspective.* Paper presented at the Spectrum Licensing and Spectrum Commons—Where to Draw the Line conference, University of Southern California, Los Angeles.

Chadderton, C. (2009). *Discourses of Britishness, race and difference. Minority ethnic students' shifting perspectives of their school experience* (Unpublished PhD thesis). Manchester, UK: Manchester Metropolitan University.

Chadderton, C. (2012). UK secondary schools under surveillance: What are the implications for race? A critical race and Butlerian analysis. *Journal of Critical Education Policy Studies, 10*(1), Retrieved from: http://www.jceps.com/PDFs/10-1-06.pdf

Chadderton, C., & Colley, H. (2012). School-to-Work transition services: Marginalising 'disposable' youth in a state of exception? *Discourse: Studies in the Cultural Politics of Education, 33*(3), 329–343.

Chisman, J. A. (1989). Apple uses simulation to improve PCB/FMS line design and operation. *Industrial Engineering, 21*(7), 40–41.

Choong, K. Y. (n.d.). A Green Revolution—Southeast Asia. Retrieved from www.berkshirepublishing. com/rvw/015/015smpl2.htm

Coase, R., Meckling, W. H., & Minasian, J. (1995). *Problems of radio frequency allocation*, DRU-1219-RC, RAND. Retrieved from http://www.rand.org/pubs/drafts/2008/DRU1219.pdf

Coase, R. H. (October, 1959). The Federal Communications Commission. *Journal of Law and Economics, 2.* 1–40. Retrieved from http://old.ccer.edu.cn/download/7874-1.pdf

Cobbett, E. (2011, September 8–9). *Elderly at the borders: The case of the South African older person's grant.* Paper presented to the Confronting the Global conference, University of Warwick, UK.

Cockburn, A. (1995). A short, meat-orientated history of the world from Eden to the Mattole. In S. Coe (Ed.), *Dead meat* (pp. 5–36). New York, NY: Four Walls, Eight Windows.

Cohen, S. (1973). *Folk devils and moral panics. Creation of mods and rockers.* London, UK: Paladin.

Colasanti, J. (2009, May 18). Pupils walk out of lessons in protest against big brother cameras. *Waltham Forest Guardian.* Retrieved from http://www.guardianseries.co.uk/news/4377621. LOUGHTON__Pupils_walk_out_of_lessons_in_protest_against_Big_Brother_cameras/

Collier, P. (2008). *The bottom billion: Why the poorest countries are failing and what can be done about it.* Oxford, UK: Oxford University Press.

Colline, D., & Phillips, T. (2011, July 15). Highway paved with good intentions drives fears of crime and destruction. *The Observer,* p. 16.

Commission on Intellectual Property Rights. (2002). *Integrating intellectual property rights and development policy.* Report of the Commission on Intellectual Property Rights. London, UK.

Communities and Small Scale Mining (2011). Retrieved from http://www.artisanalmining.org/ casm/mappopulation

Communities and Small Scale Mining (2012). Retrieved from http://artisianalmining.org

Compassion in World Farming (CIWF). (2002). *Detrimental impacts of industrial animal agriculture.* Godalming, Surrey, UK: Author.

Compassion in World Farming (CIWF). (2009). *Fact sheet—Pigs.* Retrieved from http://www.ciwf. org.uk/includes/documents/cm_docs/2010/f/factsheet_pigs.pdf

Compassion in World Farming (CIWF). (2010). *Fact sheet—Meat chickens.* Retrieved from http:// www.ciwf.org.uk/includes/documents/cm_docs/2010/f/factsheet_meat_chickens.pdf

Consultant Value Added—MMC Group Strategy Consultants. (2011). *Mobile as a trillion-dollar Industry? Check this.* Retrieved from http://consultantvalueadded.com/2011/03/07/mobile-as-a-trillion-dollar-industry-check-this/

Couldry, N. (2000) *The place of media power: Pilgrims and witnesses of the media age,* London: Routledge.

Cowhey, P. F., & Aronson, J. D., with Abelson, D. (2010). *Transforming global information and communication markets: The political economy of information.* Cambridge, MA: The MIT Press.

Critcher, C. (1993). Structures, cultures, and biographies. In S. Hall & T. Jefferson (Eds.), *Resistance through rituals: Youth subcultures in post-war Britain* (pp. 139–144). London, UK: Routledge.

Cronon, W. (1991). *Natures metropolis: Chicago and the great west.* London, UK: Norton.

Crouch, C. (1999). Employment, industrial relations and social policy. *Social Policy and Administration, 33*(4), 437–457.

Daar, A. S., Thorsteinsdóttir, H., Martin, D. K., Smith, A. C., Nast, S., & Singer, P. A. (2002, October). Top ten biotechnologies for improving health in developing countries. *Nature Genetics, 32,* 229–232.

Davenport, T. H. (1998). Putting the enterprise into the enterprise system. *Harvard Business Review*, *76*(4), 121–131.

Davis, M., & Flowers, P. (2009). Myth and HIV medical technologies: Perspectives from the 'transitions in HIV project.' In J. Burnett, P. Senker, & K. Walker (Eds.), *The myths of technology: Innovation and inequality* (pp. 145-157). New York, NY: Peter Lang.

Davison, R., Vogel, D., Harris, R., & Jones, N. (2000). Technology leapfrogging in developing countries—An inevitable luxury? *The Electronic Journal on Information Systems in Developing Countries*, *1*(5), 1–10.

Dean, J. (2007, August 11). The forbidden city of Terry Gou. *The Wall Street Journal*.

Dean, M. (2002). Liberal government and authoritarianism. *Economy and Society*, *31*(1), 37–61.

Dean, M. (2008). Governing society. *Journal of Cultural Economy*, *1*(1), 25–38.

Deleuze, G., & Guattari, F. (1988). *A thousand plateaus: Capitalism & schizophrenia* (B. Massumi, Trans.). London, UK: The Athlone Press.

de Miranda, A. (2009). Technological determinism and ideology: Questioning the 'information society' and the 'digital divide.' In J. Burnett, P. Senker, & K. Walker (Eds.), *The myths of technology: Innovation and inequality* (pp. 23–37). New York, NY: Peter Lang.

De Rivero, O. (2001). *The myth of development: Non-viable economies of the 21st century*. London, UK: Zed Books.

De Schutter, O. (2009, December). *Agribusiness and the right to food*. Geneva, Switzerland: Human Rights Council, United Nations.

DfEE. (1997). *Connecting the learning society, national grid for learning*. Retrieved from http://www.education.gov.uk/consultations/downloadableDocs/42_1.pdf

Douglas, J. (2009). Disappearing citizenship: Surveillance and the state of exception. *Surveillance & Society*, *6*(1), 32–42.

Doyle, J. (2009). Climate action and environmental activism: The role of environmental NGOs and grassroots movements in the global politics of climate change. In T. Boyce & J. Lewis (Eds.), *Climate change and the media* (pp. 103–116). London, UK: Peter Lang.

Ducombe, R., & Heeks, R. (2002). Information, ICTs and ethical trade: Implications for self-regulation (Working Paper No. 41). Manchester, UK: Centre on Regulation and Competition, Institute for Development Policy and Management, University of Manchester.

Easterly, W. (2005, March 13). A modest proposal. *The Washington Post*.

Ehrenreich, B. (2001). *Nickel and dimed: Undercover in low-wage USA*. London, UK: Granta.

Eisnitz, G. (1997). *Slaughterhouse: The shocking tales of greed, neglect and inhumane treatment inside the U.S. meat industry*. New York, NY: Prometheus Books.

England, K. & Ward, K. (eds) 'Introduction: reading neoliberalizations', in K. England and K. Ward (eds) *Neoliberalizations: States, Networks, People*. Oxford: Blackwell, pp.1-22

Escobar, A. (1995). *Encountering development: The making and unmaking of the third world*. Princeton NJ: Princeton University Press.

ESIB. (2001). ESIB statement on commodification of education. Retrieved from http://www.aic.lv/bolona/GATS/commo_statemESIB.pdf

ESIB. (2002). *Globalisation and higher education*. Policy paper on commodification of education. Paris, France: UNESCO. Retrieved from http://portal.unesco.org/education/en/files/7334/10342651350ref3_esib.doc/ref3_esib.doc

ESIB. (2005). *Policy paper on 'commodification of education.'* Bruxelles, Belgium: European Students' Union. Retrieved from http://www.esu-online.org/news/article/6064/90/

Evans, D. L. (1998). *A critical examination of claims concerning the 'impact' of print*. Retrieved from http://www.aber.ac.uk/media/Students/dle9701.html

Evans, D., & Jackson, T. (2008). *Sustainable consumption: Perspectives from social and cultural theory* (RESOLVE Working Paper No. 05-08). Guildford, Surrey, UK: University of Surrey.

Faber, D. (1993). *Environment under fire: Imperialism and the ecological crisis in Central America*. New York, NY: Monthly Review Press.

Fairtrade Foundation. (2011). *Fairtrade and fairmined gold standards launched.* Retrieved from http://www.fairtrade.org.uk/press_office/press_releases_and_statements/march_2010/fairtrade_and_fairmined_gold_standards_launched.aspx

Faulhaber, G. R. (2005). *The question of Spectrum technology, management and regime change.* Retrieved from http://www.broadbandcity.gr/content/modules/downloads/The_Question_Of_Spectrum_Technology_Management_And_Regime_Change_(Faulhaber).pdf

Feenberg, A. (1999). *Questioning technology.* London, UK: Routledge.

Feenberg, A. (2002). *Transforming technology: A critical theory revisited* (2nd ed.). New York, NY: Oxford University Press.

Fine, B. (2002). *The world of consumption: The material and cultural revisited* (2nd ed.). London, UK: Routledge.

Fine, B., Heasman, M., & Wright, J. (1996). *Consumption in the age of affluence: The world of food.* London, UK: Routledge.

Flemming, S. (2003). The leapfrog effect: Information needs for developing nations. In S. Kamel (Ed.), *Managing globally with information technology* (pp. 127–139). Hershey, PA: Idea Group.

Foresight. (2011). *The future of food and farming* (Final Project Report). London, UK: The Government Office for Science.

Foucault, M. (1991). *Discipline and punish: The birth of the prison [Surveilleret Punir]* (A. Sheridan, Trans.). London, UK: Penguin. (Original work published 1977)

Fouilhoux, M. (2004, July 12). *What GATS means to higher education.* Spain: Universidade de Santiago de Compostela. Retrieved from http://firgoa.usc.es/drupal/node/6431

Franklin, A. (1999). *Animals and modern cultures: A sociology of human-animal relations in modernity.* London, UK: Sage.

Gamble, A. (1994). *The free economy and the strong state. The politics of Thatcherism* (2nd ed.). Basingstoke, UK: Palgrave Macmillan.

Garner, R. (2008, July 30). Minister: Every school can have a police officer. *The Independent.* Retrieved from http://www.independent.co.uk/news/education/education-news/minister-every-school-can-have-a-police-officer-880315.html

Gavin, N. (2009). The web and climate change politics. In T. Boyce & J. Lewis (Eds.), *Climate change and the media* (pp. 129–144). London, UK: Peter Lang.

Gellatley, J. (1994). *The silent ark: A chilling exposé of meat—The global killer.* London, UK: Thorsons.

George, S. (1976). *How the other half dies: The real reasons for world hunger.* London, UK: Penguin.

Gibson, N., Holland, S. P., & Light, B. (1999). Enterprise resource planning: A business approach to systems development. *IEEE Proceedings of the 32nd HAWAII International Conference on System Sciences*: 1-9.

Giddens, A. (1990). *The consequences of modernity.* Cambridge, UK: Polity Press.

Giddens, A. (1998). *The third way: The renewal of social democracy.* Cambridge, UK: Polity Press.

Giddens, A. (Ed.). (2001). *The global third way debate.* Cambridge, UK: Polity Press.

Giles, J. (2009, February 10). Eating less meat could cut climate costs. *New Scientist.* Retrieved from http://www.newscientist.com/article/dn16573-eating-less-meat-could-cut-climate-costs.html.

Gillborn, D. (2006). Rethinking white supremacy: Who counts in 'whiteworld.' *Ethnicities, 6*(3), 318–340.

Gillmor, D. (2006). *We the media: Grassroots journalism by the people, for the people.* Sebastopol, CA: O'Reilly.

Giroux, H. A. (2009) *Youth in a suspect society. Democracy or disposability?* New York: Palgrave Macmillan.

Global Health Watch. (2005). *An alternative World Health Report 2005–2006.* London, UK: Zed Books.

Global Value Chains Initiative. (2012). Retrieved from http://www.globalvaluechains.org/aboutus.html

Gold Corp Inc. (2010). *Annual report.* Vancouver, British Columbia, Canada: Author.

Goldenberg, S. (2012, April 13). Eat less meat to prevent climate disaster, study warns. *The Guardian*. Retrieved from http://www.guardian.co.uk/environment/2012/apr/13/less-meat-prevent-climate-change

Graham, S. (2011). *Cities under siege. The new military urbanism*. London, UK: Verso.

Grandin, G. (2009). *Fordlandia: The rise and fall of Henry Ford's forgotten jungle city*. London, UK: Icon Books.

Gulati, S. (2008). Technology-enhanced learning in developing nations: A review. *The International Review of Research in Open and Distance Learning, 9*(1). Retrieved from http://www.irrodl.org/index.php/irrodl/article/view/477/1012

Hafner, K. (2002, May 2). Lessons learned at dot-com U. *The New York Times*. Retrieved from http://www.nytimes.com/2002/05/02/technology/lessons-learned-at-dot-com-u.html

Hammer, M. R., Bennet, M. J., & Wiseman, R. (2003). Measuring intercultural sensitivity: The intercultural development inventory. *International Journal of Intercultural Relations, 27*(4), 421–443.

Hara, N., & Kling, R. (2001). Student distress in web-based distance education. *EDUCAUSE Quarterly, 3*, 68–69. Retrieved from http://net.educause.edu/ir/library/pdf/eqm01312.pdf

Haraway, D. J. (2008). *When species meet*. Minneapolis: University of Minnesota Press.

Harmony Gold Mining Company Limited. (2010). *Annual report*. Randfontein, South Africa: Author.

Harris, M. (1987). *The sacred cow and the abominable pig*. New York, NY: Touchstone/Simon and Shuster.

Harris, R., Kumar, A., & Balaji, V. (2003). Sustainable telecentres? Two cases from India. In S. Krishnaa & S. Maldon (Ed.), *The digital challenge: Information technology in the development context* (pp. 124–135). Aldershot, UK: Hants, Ashgate.

Harvey, D. (2003). *The new imperialism*. Oxford, UK: Oxford University Press.

Harvey, D. (2005). *A brief history of neoliberalism*. Oxford, UK: Oxford University Press.

Harvey, D. (2006). *The limits to capital* (2nd ed.). London, UK: Verso.

Hauben, M. (1995). Participatory democracy from the 1960s and SDS into the future on-line. Retrieved from http://www.columbia.edu/~hauben/CS/netdemocracy-60s.txt

Hawari, A., & Heeks, R. (2010). *Explaining ERP failure in developing countries: A Jordanian case study* (Working Paper Series No. 45). Manchester, UK: Centre for Development Informatics, Institute for Development Policy and Management, Manchester University.

Hayek, F. (1960). *The constitution of liberty*. London, UK: Routledge & Kegan Paul.

Hazlett, T. W. (2001a, March 28). *Clock ticking for Spectrum. CNET News*. Retrieved from http://news.cnet.com/Clock-ticking-for-spectrum/2010-1071_3-281424.html

Hazlett, T. W. (2001b). The wireless craze, the unlimited bandwidth myth, the Spectrum auction faux pas, and the punchline to Ronald Coase's 'big joke.' *Harvard Journal of Law & Technology, 14*(2), 337–567.

Hazlett, T. W., Porter, D., & Smith, V. (2009). *Radio Spectrum and the disruptive clarity of Ronald Coase* (George Mason Law & Economics Research Paper No. 10-18). Retrieved from http://www.iep.gmu.edu/documents/Draft.TWH.11.9.23.Z.pdf

Head, S. (2003). *The new ruthless economy—Work and power in the new digital age*. New York, NY: Oxford University Press.

Heaford, J. M. (1983). *Myth of the learning machine: The theory and practice of computer based training*. Wilmslow, UK: Sigma Technical Press.

Heath, J., & Potter, A. (2004). *Nation of rebels: Why counterculture became consumer culture*. New York, NY: HarperCollins.

Heeks, R. (2002). Information systems and developing countries: Failure, success, and local improvisations. *The Information Society, 18*(2), 101–112.

Held, D. (2004). *Global covenant: The social Democratic alternative to the Washington consensus*. Cambridge, UK: Polity Press.

Hine, R. V., & Faragher, J. M. (2000). *The American west: A new interpretive history*. New Haven, CT: Yale University Press.

Hirschfield, P. (2009). School surveillance in America. Disparate and unequal. In T. Monahan & R. D. Torres (Eds.), *Schools under surveillance. Cultures of control in public education* (pp. 38–54). New Brunswick, NJ: Rutgers University Press.

Hobday, M. (1994). Technological learning in Singapore: A test case of leapfrogging. *The Journal of Development Studies, 30*(3), 831–858.

Hoeckman, B. M., Maskus, K. E., & Saggi, K. (2004). *Transfer of technology to developing countries* (Working Paper PEC2004-0003). Boulder: Institute of Behavioural Science, University of Colorado.

Hollow, D. (2009). *eLearning in Africa: Challenges, priorities and future direction.* Retrieved from http://www.gg.rhul.ac.uk/ict4d/workingpapers/Hollowelearning.pdf

Hope, A. (2009). CCTV, school surveillance and social control. *British Educational Research Journal, 35*(6), 891–907.

Hope, A. (2010). Seductions of risk, social control and resistance to school surveillance. In T. Monahan & R. D. Torres (Eds.), *Schools under surveillance: Cultures of control in public education* (pp. 230–246). New Brunswick, NJ: Rutgers University Press.

Hudson, H. (2001). The potential of ICT'S for development: Opportunities and obstacles (Background Paper for the *World Employment Report*). Geneva, Switzerland: International Labour Organization.

Humphrey, J., & Schmitz, H. (2001). *Governance in global value chains.* Brighton, Sussex, UK: Institute of Development Studies.

Hurwitz, J. (1998). Customizing packaged applications. *DBMS, 11*(4), 10–12. Retrieved from http://dl.acm.org/citation.cfm?id=298045&picked=prox&CFID=124867924&CFTOKEN=65558484

IntoMobile. (2010). *India 3G winners announced.* Retrieved from http://www.intomobile.com/2010/05/22/india-3g-winners-announced/

ITU. (2010). *Measuring the information society.* Retrieved from http://www.itu.int/ITU-D/ict/publications/idi/index.html

Jackson, T. (2011). *Prosperity without growth: Economics for a finite planet.* London, UK: Earthscan.

Jessop, B. (2002). *The future of the capitalist state.* Cambridge, UK: Polity Press.

Johnson, A. (1991). *Factory farming.* Oxford, UK: Blackwell.

Jordan, T. G. (1993). *North American cattle ranching frontiers.* Albuquerque: University of New Mexico Press.

Jowell, T. (2002, May 7). Culture secretary statement to parliament on presentation of the draft Communications Bill. *BBC News.* Retrieved from http://news.bbc.co.uk/1/hi/uk_politics/1973563.stm

Judge, P. (2004, November). Ofcom to throw Radio Spectrum wide open. *Techworld, 23.* Retrieved from http://features.techworld.com/mobile-wireless/1013/ofcom-to-throw-radio-spectrum-wide-open/

Judge, P. (2010). UK Radio Spectrum auctions hang in the balance. *Tech Week Europe.* Retrieved from http://www.eweekeurope.co.uk/news/news-it-infrastructure/uk-radio-spectrum-auctions-hang-in-the-balance-3145

Kahn, R., & Kellner, D. (2003). Internet subcultures and oppositional politics. In D. Muggleton & R. Weinzierl (Eds.), *The post-subcultures reader* (pp. 294–309). London, UK: Berg.

Kamel, S. (2010). Information and communications technology for development: Building the knowledge society in Egypt. In L. B. Shaver & N. Rizk (Eds.), *Access to knowledge in Egypt: New research on intellectual property, innovation and development* (pp. 174–204). London, UK: Bloomsbury Academic.

Kamhawi, E. M. (2008). Enterprise resource-planning systems adoption in Bahrain: Motives, benefits, and barriers. *Journal of Enterprise Information Management, 21*(3), 310–334.

Kannan, K. P. (2003). India's tenth plan and Kerala's development challenges. *Kerala Calling, 24*(12), 3–5 and 36–38.

Kannan, P., & Vijayamohanan Pillai, N. (2004, August). *Development as a right to freedom: An interpretation of the 'Kerala model'* (pp. 27–33). Trivandrum, Kerala India: Centre for Development Studies.

Kawamoto, K. (2003). *Digital journalism. Emerging media and the changing horizons of journalism.* New York, NY: Rowman & Littlefield.

Khallili, M. (2011, September 5). Why I rioted: One man's personal take. *The Guardian.* Retrieved from http://www.guardian.co.uk/uk/video/2011/sep/05/why-i-rioted-video?intcmp=239

Kiely, R. (2007). *The new political economy of development: Globalisation, imperialism, hegemony.* Basingstoke, UK: Palgrave.

Kiggundu, M. (1989). *Managing organizations in developing countries: An organizational and strategic approach.* West Hartford, CT: Kumerian Press.

Kingsley, P. (2011, November 2). Despair and desperation—The real story of youth unemployment in Britain. *Guardian, G2.*

Kingsnorth, P. (2003). *One no, many yeses: A journey to the heart of the global resistance movement.* New York, NY: Free Press.

Kiraka, R. N., & Manning, K. (2002). Getting online: Australian international development agencies and ICT use. *Journal of International Development, 14*(1), 75–87.

Klaassen, C. D. (Ed.). (2008). *Casarett and Doull's toxicology.* New York, NY: McGraw-Hill Medical.

Klein, J., Reynolds, M., & Johnston, D. (2007). *Advice on spectrum usage, HDTV and MPEG-4: Part of the HDTV public value test.* Retrieved from http://www.bbc.co.uk/bbctrust/assets/files/pdf/consult/hdtv/sagentia.pdf

Kothari, U., & Minogue, M. (Ed.). (2002). *Development theory and practice: Critical perspectives.* Basingstoke, UK: Palgrave.

Kropiwnicka, M. (2005). Biotechnology and food security in developing countries: The case for strengthening international environmental regimes. *ISYP Journal on Science and World Affairs, 1,* 45–60.

Kumar, K., & Hillegersberg, V. J. (2000). ERP experiences and evolution. *Communications of the ACM, 43*(4), 23–26.

Kumar, R., & Best, M. L. (2006). Impact and sustainability of e-government services in developing countries: Lessons learned from Tamil Nadu, India. *The Information Society, 22*(1), 1–12.

Kupchik, A., & Bracy, N. L. (2009). To protect, serve and mentor? Police officers in public schools. In T. Monahan & R. D. Torres (Eds.), *Schools under surveillance Cultures of control in public education* (pp. 21–37). New Brunswick, NJ: Rutgers University Press.

Labaton, S. (2003, May 16). US moves to allow trading of Radio Spectrum licenses. *The New York Times.* Retrieved from http://www.nytimes.com/2003/05/16/business/us-moves-to-allow-trading-of-radio-spectrum-licenses.html

Lammy, D. (2011). David Lammy MP reacts to the Tottenham riots. Retrieved from http://www.youtube.com/watch?v=DVEQFsjY7pY

Langstone, D. (2009). Myths, crimes and videotape. In J. Burnett, P. Senker, & K. Walker (Eds.), *The myths of technology: Innovation and inequality* (pp. 113–127). New York, NY: Peter Lang.

Lappé, F. M., & Collins, J. (1982). *Food first: Beyond the myth of scarcity.* London, UK: Abacus.

Larmer, B. (2009, January). The price of gold. *National Geographic, 251*(1), 34–61.

Lawrence, F. (2011, November 24). Fat profits health hangover as big brands woo world's poorest shoppers. *The Guardian.*

Layard, R. (2005). *Happiness: Lessons from a new science.* London, UK: Allen Lane.

LeDuff, C. (2003). 'At the slaughterhouse, some things never die.' In C. Wolfe (Ed.), *Zoontologies: The question of the animal.* Minneapolis: University of Minnesota Press, (pp. 183–198) (Reprinted from, *New York Times,* June 16, 2000, available at http://www.nytimes.com/library/national/race/061600leduff-meat.html).

Levidow, L. (2002). Marketising higher education: Neoliberal strategies and counter-strategies. In K. Robins & F. Webster (Eds.), *The virtual university? Knowledge, markets and management* (pp. 227–248). Oxford, UK: Oxford University Press.

Lewis, P., Ball, J., & Taylor, M. (2011, September 5). Riot jail sentences in crown courts longer than normal. *The Guardian*. Retrieved from http://www.guardian.co.uk/uk/2011/sep/05/riot-jail-sentences-crowcourts?intcmp=239

Lines, T. (2008). *Making poverty: A history*. London, UK: Zed Books.

Lister, R. (1998). From equality to social exclusion: New labour and the welfare state. *Critical Social Policy*, *18*(55), 215–225.

Lovink, G. (2002). *Dark fiber: Tracking critical internet culture*. Cambridge, MA: The MIT Press.

Lübker, M. (2007) "Labour Shares", an ILO Policy Brief, Geneva:International Labour Organisation http://www.ilo.org/wcmsp5/groups/public/---dgreports/---integration/documents/publication/wcms_086237.pdf accessed 14 February 2007

Lyon, D. (Ed.). (2002). *Surveillance as social sorting: Privacy, risk, and digital discrimination*. London, UK: Routledge.

Mabert, V. A., Soni, A., & Venkataramanan, M. A. (2003). Enterprise resource planning: Managing the implementation process. *European Journal of Operational Research*, *146* (2)30–314.

MacDonald, M. (2010, March/April). Eat like it matters, footprints in the future of food. *Resurgence*, *259*, 32–33.

Magnussen, J., Vrangbaek, K., & Saltman, R. B. (Eds.). (2009). *Nordic health care systems: Recent reforms and current policy challenges*. Maidenhead, UK: Open University Press.

MailOnline. (2006, August 18). The stark reality of iPod's Chinese factories. *Daily Mail*. Retrieved from http://www.dailymail.co.uk/news/article-401234/The-stark-reality-iPods-Chinese-factories.html

Mansell, R., & Silverston, R. (Eds.). (1996). *Communication by design. The politics of information and communication technologies*. Oxford, UK: Oxford University Press.

Mansell, R., & Wehn, U. (1998). *Knowledge societies: Information technology for sustainable development*. Oxford, UK: Oxford University Press.

Marcus, E. (2005). *Meat market: Animals, ethics and money*. Boston, MA: Brio Press.

Markoff, J. (2005). *What the doormouse said—How the sixties counterculture shaped the personal computer industry*. London, UK: Penguin Books.

Marx, G., & Steeves, V. (2010). From the beginning: Children as subjects and agents of surveillance. *Surveillance & Society*, *7*(3/4), 192–230.

Marx, K. (1976). *Capital vol. 1*. London, UK: Penguin Books. (Original work published 1867)

Marx, K. (1981). *Capital vol. 3*. London, UK: Penguin Books. (Original work published 1894)

Marx, K. (1998). *Capital vol. 1*. London, UK: Electronic book Company.

Mason, J., & Finelli, M. (2006). Brave new farm? In P. Singer (Ed.), *In defense of animals: The second wave*. Oxford, UK: Blackwell.

Masson, J. M. (2004). *The pig who sang to the moon: The emotional world of farm animals*. London, UK: Jonathan Cape.

McCahill, M., & Finn, R. (2010). The social impact of surveillance in three UK schools: 'Angels', 'devils' and 'teen mums.' *Surveillance & Society*, *7*(3/4), 273–289.

McCauley, D. (2004). *Reaping the benefits of ICT: Europe's productivity challenge*. London, UK: The Economist Intelligence Unit.

McDougall, D. (2010, September 12). Child miners slave for gold in mud and poison. *The Sunday Times*. Retrieved from http://www.thesundaytimes.co.uk/sto/news/world_news/Africa/article392477.ece

McGovern, G. (2003, October 20). Why personalization hasn't worked. *New Thinking*. Retrieved from http://www.wissensnavigator.com/documents/Personalization.pdf

McGreal, C. (2012, January 10). The US schools with their own police. *The Guardian*. Retrieved from: http://www.guardian.co.uk/world/2012/jan/09/texas-police-schools

McKenna, R. (1991). *Relationship marketing: Successful strategies for the age of the customer*. Reading, MA: Addison-Wesley.

McLoughlin, J., Rosen, P., Skinner, D., & Webster, A. (1999). *Valuing technology*. London, UK: Routledge.

Mehmet, O. (1999). *Westernizing the third world: The Eurocentricity of economic development theories.* London, UK: Routledge.

Mendick, R., Murphy, J., & Low, V. (2008, February 27). The graduate eco-warrior commons raid. *London Evening Standard,* pp. 1–3.

Milcent, C. (2005). *Hospital ownership, reimbursement systems and mortality.* Retrieved from http://www.econ.puc-rio.br/pdf/seminario/2005/carine.pdf

Miliband, E. (2008, January 30). *Fighting poverty and inequality in an age of affluence: 100 years on from the Poor Law Minority Report.* London: Commonwealth Club.

Ministry of Agriculture, Fisheries and Food (MAFF). (1991, August). *The fresh meat and poultry meat (hygiene, inspection, and examinations for residues) (charges) regulations 1990* (Circular FSH 1/91). Surbiton, UK.

Minton, A. (2009). *Ground control.* London, UK: Penguin.

Misund, G., & Hoiberg, J. (2003). Sustainable information technology for global Sustainability, in *Information Resources for Global Sustainability: Proceedings of the 3rd International Symposium on Digital Earth,* Brno, Czech Republic.

Molla, A., & Bhalla, A. (2006). Business transformations through ERP: A case study of an Asian company. *Journal of Information Technology Case and Application Research, 8*(1), 34–54.

Monahan T., & Fisher, J. A. (2008). Editorial: Surveillance and inequality. *Surveillance & Society, 5*(3), 217–226.

Monahan, T., & Torres, R. D. (2010). Introduction. In T. Monahan & R. D. Torres (Eds.), *Schools under surveillance. Cultures of control in public education* (pp. 1–18). New Brunswick, NJ: Rutgers University Press.

Moore G. E. (1965), Cramming more components onto integrated circuits, *Electronics Magazine 38*(8), April 19. Retrieved from http://download.intel.com/museum/Moores_Law/ArticlesPress_Releases/Gordon_Moore_1965_Article.pdf

Moore, G. E. (1975). Progress in digital integrated electronics. *Electron Devices Meeting 1975 International,* IEEE, pp. 11–13.

Morozov, E. (2011). *The net delusion: How not to liberate the world* (Kindle ed.). New York, NY: Penguin.

Murphy, J. (2008, February 27). The storming of Parliament. *London Lite, 1-3.*

Musungu, S., & Oh, C. (2006). The use of flexibilities in TRIPS by developing countries: Can they promote access to medicines?" Commission on Intellectual Property Rights, Innovation and Public Health, World Health Organization, Geneva.

Neate, R. (2008, October 6). Steve Wozniak interview. *Daily Telegraph.* Retrieved from http://www.telegraph.co.uk/finance/newsbysector/mediatechnologyandtelecoms/3145691/Steve-Wozniak-interview-iconic-co-founder-on-the-iPod-iPhone-and-future-for-Apple

Newlands, M. (2010). *Come together: Professional practice and radical protest.* Retrieved from http://www.proof-reading.org/come-together-professional-practice-and-radical-protest

Newmont Mining Corporation. (2010). *Annual report.* Greenwood Village, CO: Newmont Mining Corporation.

Nibert, D. (2002). *Animal rights/human rights: Entanglements of oppression and liberation.* Lanham, MD: Rowman & Littlefield.

Nightingale, P., & Martin, P. (2009). The myth of the biotech revolution. In J. Burnett, P. Senker, & K. Walker (Eds.), *The myths of technology: Innovation and inequality.* New York, NY: Peter Lang.

Noble, D. F. (1997). *Digital diploma mills, part I: The automation of higher education.* Retrieved from http://communication.ucsd.edu/dl/ddm1.html

Nokia. (2010). *Materials.* Retrieved from http://www.nokia.com/global/about-nokia/people-and-planet/sustainable-devices/materials/materials/

Nokia. (2011). *Sustainability report 2011.* Retrieved from http://stream.heartshapedwork.com/2011/07/01/nokia-sustainability-report-2011/

Notes from Nowhere. (2003). *We are everywhere: The irresistible rise of global anticapitalism.* London, UK: Verso.

OECD. (2008). *Growing unequal? Income distribution and poverty in OECD countries.* Paris, France: Author.

Ofcom. (2003).*Consultation on spectrum trading: A summary.* London, UK: Author.

Ofcom. (2004). *Spectrum trading review.* London, UK: Author.

Ofcom (2005) Spectrum Trading Review: Implementation Plan, published 13.1.2005. Retrieved from http://stakeholders.ofcom.org.uk/binaries/consultations/sfrip/summary/sfr-plan.pdf

Ofcom. (2007, November). *Ofcom consultation guidelines.* Retrieved from http://stakeholders.ofcom.org.uk/consultations/how-will-ofcom-consult

Ofcom. (2008). *The wireless telegraphy (spectrum trading) (amendment) (no. 2) regulations 2008* (SI 2008, No. 2105 Electronic Communications). Retrieved from http://www.legislation.gov.uk/uksi/2008/2105/regulation/1/made

Ofcom. (2009, September 22). *Simplifying spectrum trading: Regulatory reform of the spectrum trading process and introduction of spectrum leasing—Consultation.* Retrieved from http://stakeholders.ofcom.org.uk/binaries/consultations/simplify/summary/simplify.pdf

Ofcom. (2010). *Simplifying spectrum trading: Reforming the spectrum trading process and introducing spectrum leasing. Interim statement.* Retrieved from http://stakeholders.ofcom.org.uk/binaries/consultations/simplify/statement/statement.pdf

Ofcom. (2011a, March 22). *Consultation on assessment of future mobile competition and proposals for the award of 800 mhz and 2.6 ghz spectrum and related issues.* Retrieved from http://stakeholders.ofcom.org.uk/binaries/consultations/combined-award/summary/combined-award.pdf

Ofcom. (2011b, June 2). *Coexistence of new services in the 800 mhz band with digital terrestrial television.* Retrieved from http://stakeholders.ofcom.org.uk/consultations/coexistence-with-dtt/

Ofcom. (2011c, August 4). *A nation addicted to smart phones.* Retrieved from http://media.ofcom.org.uk/2011/08/04/a-nation-addicted-to-smartphones/

Office for National Statistics (ONS). (2011). *Labour market statistics October: A06 educational status and labour market status (employment, unemployment and inactivity) of people aged from 16 to 24.* Retrieved from http://www.ons.gov.uk/ons/publications/re-reference-tables.html?edition=tcm%3A77-222441

Osler, A., & Starkey, H. (2005). *Changing citizenship: Democracy and inclusion in education.* Buckingham, UK: Open University Press.

O'Toole, K. (2001, February 1). Going once, going twice . . . SOLD on auctions. *Stanford Graduate School of Business (GSB) News.* Retrieved from http://www.gsb.stanford.edu/news/research/gametheory_auctions.shtml?cmpid=researchECON

O'Toole, T. (2007, September 27). *Teaching race and ethnicity in practice: Race and citizenship education.* Paper presented at the Higher Education Academy, University of Birmingham, UK.

Parayil, G. (2005). The digital divide and increasing returns: Contradictions of informational capitalism. *The Information Society, 21*(1), 41–51.

Parliamentary Office of Science and Technology. (2005). *Fighting diseases of developing countries* (Postnote No. 1241). London, UK.

Parliamentary Office of Science and Technology. (2006). *Food security in developing countries* (No. 06/274). London, UK.

Parsons, C. (2008). Race relations legislation, ethnicity and disproportionality in school exclusions in England. *Cambridge Journal of Education, 38*(3), 401–419.

Patel, R. (2007). *Stuffed and starved: Markets, power and the hidden battle for the world's food system.* London, UK: Portobello Books.

Payscale. (2012). Retrieved from http://www.payscale.com/research/ZA/Industry=Gold_Mining/Salary

Peet, R., & Hartwick, E. (1999). *Theories of development.* New York, NY: Guilford Press.

Pemberton, C. (2010). *Youth crime down but number of children in prison too high.* Retrieved from http://www.communitycare.co.uk/Articles/2010/05/28/114608/youth-crime-down-but-number-of-children-in-prison-too-high.htm

Pickerill, J. (2003). *Cyberprotest: Environmental activism on-line*. Manchester, UK: Manchester University Press.

Pieterse, J. N. (2002). Global inequality bringing politics back in. *Third World Quarterly, 23*(6), 1023–1046.

Pollan, M. (2003, March 31). Power steer. *The New York Times Magazine*. Retrieved from http://www.nytimes.com/2002/03/31/magazine/power-steer.html

Pollock, N., Procter, R., & Williams, R. (2003). Fitting standard software packages to non-standard organisations. *Technology Analysis and Strategic Management, 15*(3), 317–332.

Pollock, N., & Williams, R. (2008). *Software and organisations: The biography of the enterprise-wide system or how SAP conquered the world*. London, UK: Routledge.

Pollock, N., Williams, R., & D'Adderio, L. (2007). Global software and its provenance: Generification work in the production of organizational software packages. *Social Studies of Science, 37*(2), 254–280.

Pomfret, J., & Soh, K. (2010). For Apple suppliers, loose lips can sink contracts. *Reuters*. Retrieved from http://www.reuters.com/article/idUSTRE61G3XA20100217

Potter, N. (1979, May 7). A computer of one's own. *New York Magazine*, pp. 59–61.

Prahalad, C. K. (2006, Autumn). The innovation sandbox. *Strategy + Business, 44*.

Prebble, L. (2009). *Enron*. London, UK: Methuen Drama.

Race for Justice. (2006). *Race and the criminal justice system. Facts and Figures*. Retrieved from http://www.raceforjustice.net/jdd/public/documents/file/pdf%20downloads/Facts%20and%20Figures.pdf

Radio Authority. (2002, May 10). *Report of the independent review of spectrum management— Response by the Radio Authority*, p. 2. Retrieved from http://www.ofcom.org.uk/static/archive/ra/spectrum-review/comments/radio-authority.pdf

Ragnedda, M. (2010). Review of Monahan and Torres' (Eds.) *Schools under surveillance. Surveillance & Society, 7*(3/4), 356–357.

Rajapakse, J., & Seddon, P. (2005). *Why ERP may not be suitable for organisations in developing countries in Asia* (Working Paper No. 121). Australia: Department of Information Systems, University of Melbourne. Retrieved from http://www.pacis-net.org/file/2005/121.pdf

RAND. (2002). About RAND. Retrieved from http://www.rand.org/about/faq.html

Ransom, D. (1974). Ford country: Building an elite for Indonesia. In S. Weissman (Ed.), *The Trojan horse: A radical look at foreign aid*. San Francisco, CA: Ramparts Press.

Rees, M. (2001). Facing an empty victory. *Time, 158*(10), 33.

Reicher, S., & Stott, C. (2011, November 18). Mad mobs and Englishmen. *The Guardian*. Retrieved from http://www.guardian.co.uk/science/2011/nov/18/mad-mobs-englishmen-2011-riots

Rempel, W. C. (1985, March 8). Apple to shut 4 plants for week to trim stocks. *Los Angeles Times*, p. 2.

Reuters. (2010). Retrieved from http://uk.reuters.com/article/idUKTRE67N5PW2010082

Rezaie, R., & Singer, P. A. (2010). Global health or global wealth? *Nature Biotechnology, 28*(9), 907–909.

Rifkin, J. (1994). *Beyond beef: The rise and fall of cattle culture*. London, UK: Thorsons.

Rikowski, G. (2002a). *Globalisation and education*. Paper prepared for the House of Lords Select Committee on Economic Affairs—Inquiry into the Global Economy. Retrieved from http://www.leeds.ac.uk/educol/documents/00001941.htm

Rikowski, G. (2002b). Transfiguration: Globalisation, the World Trade Organisation and the national faces of the GATS. Retrieved from http://firgoa.usc.es/drupal/files/rikowski-transfiguration-gats.pdf

Rikowski, G. (2007). The commodification of education. Retrieved from http://www.flowideas.co.uk/?page=articles&sub=The%20Commodification%20of%20Education

Ritvo, H. (1990). *The animal estate: The English and other creatures in the Victorian age*. Harmondsworth, UK: Penguin.

Ritzer, G. (2004). *The globalization of nothing*. London: Sage/Pine Forge Press.

Roberts, P. (2008). *The end of food: The coming crisis in world food industry*. London, UK: Bloomsbury.

Robbins, D. (2012, January). After Texas school shooting, many questions loom. *The Guardian, 6*. Retrieved from http://www.guardian.co.uk/world/feedarticle/10027333

Rogers, B. (2004). *Beef and liberty: Roast beef, John Bull and the English nation*. London, UK: Vintage.

Roszak, T. (2000). From Satori to Silicon Valley: San Francisco and the American counterculture. Retrieved from http://library.stanford.edu/mac/primary/docs/satori/down.html

Ryan, Ó. (2011). *Chocolate nations: Living and dying for cocoa in West Africa*. London, UK: Zed Books.

Sachs, J. (2005). *The end of poverty: How we can make it happen in our lifetime*. London, UK: Penguin.

Saltman, K. J., & Gabbard, D. A. (2011). Introduction to the second edition. In K. J. Saltman & D. A. Gabbard (Eds.), *Education as enforcement* (2nd ed., pp. 19–26.). New York, NY: Routledge.

Samuels, M. (2005, January). Lessons to be learned from failure of UKeU. *Computing, 20*. Retrieved from http://www.computing.co.uk/ctg/analysis/1839918/lessons-learned-failure-ukeu

Sapp, G., & Jones, J. (2000, March 27). Bandwidth exchanges due: Emerging companies look to broker network lifeblood. *InfoWorld*. Retrieved from http://www.accessmylibrary.com/article-1G1-60805772/bandwidth-exchanges-due-emerging.html

Saunders, P. (2010). *Beware false prophets: Equality, the good society and the spirit level*. London, UK: Policy Exchange.

Schech, S. (2002). Wired for change: The links between ICTs and development discourses. *Journal of International Development, 14*(1), 13–23.

Scherrer, C. (2005). GATS: Long-term strategy for the commodification of education. *Review of International Political Economy, 12*(3), 484–510.

Schlosser, E. (2002). *Fast food nation*. London, UK: Penguin.

Schumpeter, J. A. (1954). *Capitalism, socialism and democracy* (4th ed.). London, UK: George Allen and Unwin.

Scoones, I. (2006). *Science, agriculture and the politics of policy: The case of biotechnology in India*. Hyderabad, India: Orient Longman.

Sehgal, V., Dehoff, K., & Panneer, G. (2010, Summer). The importance of frugal engineering. *Strategy + Business, 59*.

Seligson, M. A., & Passe-Smith, J. T. (2008). *Development and underdevelopment: The political economy of global inequality* (4th ed.). Boulder, CO: Lynne Rienner.

Sen, A. (1999). *Development as freedom*. Oxford, UK: Oxford University Press.

Sen, A. (2009, March 14). Capitalism beyond the crisis. *The Guardian*.

Senker, P. (1992). *Industrial training in a cold climate: An assessment of Britain's training policies*. Aldershot, UK: Avebury.

Senker, P. (2000). A dynamic perspective on technology, economic inequality and development. In S. Wyatt, F. Henwood, N. Miller, & P. Senker (Eds.), *Technology and in/equality: Questioning the information society* (pp. 197–217). London and New York: Routledge.

Senker, P. (2011). *The sad history of ISTCs in England*. Retrieved from http://petersenker.org.uk/the-sad-history-of-istcs-in-england

Senker, P., & Chataway, J. (2009). The myths of agricultural technology. In J. Burnett, P. Senker, & K. Walker (Eds.), *The myths of technology: Innovation and inequality* (pp. 171–184). New York, NY: Peter Lang.

Sennett, R. (2005). *The culture of the new capitalism*. New Haven, CT: Yale University Press.

Sharpe, R. (2009). Inequalities in the globalised knowledge-based economy. In J. Burnett, P. Senker, & K. Walker (Eds.), *The myths of technology: Innovation and inequality* (pp. 39–52). New York, NY: Peter Lang.

Sia S. K & Soh, C. (2007), An assessment of package-organisation misalignment: Institutional and ontological structure, *European Journal of Information Systems*, 16 (5): 568-583.

Sidwell, M. (2008). *Unfair trade*. London, UK: Adam Smith Institute.

Silverstone, R., & Mansell, R. (Eds.). (1996). *Communication by design: The politics of information and communication technologies*. Oxford, UK: Oxford University Press.

Simmons, L. (2010). The docile body in school space. In T. Monahan & R. D. Torres (Eds.), *Schools under surveillance. Cultures of control in public education* (pp. 55–72). New Brunswick, NJ: Rutgers University Press.

Simpson, C. (2010, December 23). Shooting gold diggers at African mine seen amid record prices. Retrieved from http://www.bloomberg.com/news/2010-12-23/shooting-gold-diggers-at-african-mine-seen-amid-record-prices.html

Sinclair, U. (1982). *The jungle*. Harmondsworth, UK: Penguin. (Original work published 1906)

Singh S. (2006). *Food security effectiveness of the public distribution system in India*. Ljubljana, Slovenia: University of Ljubljana Faculty of Economics and Center for Promotion of Enterprises.

Sinn, H-W. (2012). *The green paradox: A supply-side approach to global warming*. Cambridge, MA: The MIT Press.

Skeggs, B. (2004). *Class, self, culture*. London, UK: Routledge.

Smith, J., & Oliver, M. (1992). Automation isn't always the answer (part 8). *Machine Design, 64*(23), 45.

Smith, M. (1985). *Radio, TV and cable: A telecommunication approach*. New York, NY: CBS.

Soros, G. (2008). *The crisis of global capitalism*. London, UK: Little, Brown.

Sow, C., Kien, S. S., & Yap, T. J. (2000). Cultural fits and misfits: Is ERP a universal solution? *Communications of the ACM, 43*(4), 47–51.

Steinfeld, H., Gerber, P., Wassenaar, T., Castel, V., Rosales, M., & de Haan, C. (2006). *Livestock's long shadow*. Rome, Italy: UN FAO.

Steinmueller, E. (2001). ICTs and the possibilities for leapfrogging by developing countries. *International Labour Review, 140*(2), 193–210.

Stephens, N. (2010). In vitro meat: Zombies on the menu? Retrieved from http://www.law.ed.ac.uk/ahrc/script-ed/vol7-2/stephens.asp

Stiglitz, J. (2002). *Globalisation and its discontents*. London, UK: Allen Lane.

Sturgeon, T. J., & Kawakami, M. (2010). Global value chains in the electronics industry: Was the crisis a window of opportunity for developing countries? In O. Gatteano, G. Gereffi, & C. Staritz (Eds.), *Global value chains in a post crisis world* (pp. 245–302). Washington, DC: World Bank.

Supergreenme. (2009, June 10). Artisanal gold mining. Retrieved from http://www.supergreenme.com/go-green-environment-eco:Artisanal-Gold-Mining

Sussman, G., & Lent, J. A. (1998). *Global productions: Labor in the making of the 'information society.'* New York, NY: Hampton Press.

Sutherland, K. (2008, February 27). Protesters scale parliament. *The London Paper*, p. 3.

Sutton, M. A., Howard, C. M., Ensman, J. W., Billen, G., Bleeker, A., Grennfelt, P., van Grinsven, H., & Grizzetti, B. (Eds.). (2011). *The European nitrogen assessment: Sources, effects and policy perspectives*. Cambridge, UK: Cambridge University Press.

Tatara, K., & Okamoto, E. (2009). Japan: Health system review. *Health Systems in Transition, 11*(5), 1–164.

Taylor, E. (2009). I spy with my little eye: The use of CCTV in schools and the impact on privacy. *The Sociological Review, 58*(3), 381–405.

The Economist. (2011, January 20). Frugal healing: Inexpensive Asian innovation will transform the market for medical devices. Retrieved from http://www.economist.com/node/17963427

The Poverty Site. (2011). *United Kingdom. School exclusions*. Retrieved from http://www.poverty.org.uk/27/index.shtml

The Telegraph. (2012, May 17). 4G Networks to disrupt TV signals for two million homes. Retrieved from http://www.telegraph.co.uk/technology/mobile-phones/9271771/4G-networks-to-disrupt-TV-signals-for-two-million-homes.html

Thomas, K. (1983). *Man and the natural world: Changing attitudes in England 1500–1800*. London, UK: Allen Lane.

Thompson, J., Millstone, E., Scoones, I., Ely, A., Marshall, F., Shah, E., & Stagl, S. (2007). *Agri-Food system dynamics: Pathways to sustainability in an era of uncertainty* (STEPS Working Paper 4). Brighton, UK: STEPS Centre.

Thompson, M. P. A. (2004). ICT, power and developmental discourse: A critical analysis. *Electronic Journal of Information Systems in Developing Countries, 20*(4), 1–26.

Todaro, M. P., & Smith, S. C. (2003). *Economic development* (8th ed.). Harlow, UK: Pearson Education.

Torres, B. (2007). *Making a killing: The political economy of animal rights.* Oakland, CA: AK Press.

Toynbee, P. (2003). *Hard work: Life in low-pay Britain.* London, UK: Bloomsbury.

Travis, A. (2010, June 17). 70% rise in numbers of black and Asian people stopped and searched. *The Guardian.* Retrieved from http://www.guardian.co.uk/law/2010/jun/17/stop-and-search-police

Traxler, J. (2010, May 26–28). *Making good use of mobile phone capabilities* (eLA 2007). Paper presented at the 5th International Conference on ICT for Development, Education and Training, Accra, Ghana. Retrieved from http://www.elearning-africa.com/newsportal/english/news70_print.php#

Twerefou, D. K. (2009). *Mineral exploitation, environmental sustainability and sustainable development in EAC, SADC and ECOWAS regions* (ATPC Work in Progress No. 79). Addis Ababa, Ethiopia: African Trade Policy Centre.

Tyler, W. (2003). Dancing at the edge of chaos: A spanner in the works of global capitalism. In Notes From Nowhere (Ed.), *We are everywhere: The irresistible rise of global anticapitalism* (pp. 188–195). London, UK: Verso.

UNESCO. (2005). *UN decade of education for sustainable development 2005–2014.* Retrieved from http://unesdoc.unesco.org/images/0014/001416/141629e.pdf

UNESCO. (2008). *Quality education, equity and sustainable development: A holistic vision through UNESCO's four World Education Conferences 2008–2009.* Retrieved from http://www.scribd.com/doc/45734205/Quality-Education

United Nations. (1990). *Human development report.* Oxford, UK: Oxford University Press.

United Nations General Assembly. (2000, September 18). Resolution 2, session 55, United Nations Millennium Declaration.

United Nations, (2010), Towards a New International Development Architecture for LDCs, The Least Developed Countries Report, Available at http://unctad.org/en/docs/ldc2010_en.pdf

United Nations (2011), The Potential Role of South-South Cooperation for Inclusive and Sustainable Development, The Least Developed Countries Report, Available at http://unctad.org/en/docs/ldc2011_en.pdf

Unwin, T. (2008). *Survey of e-learning in Africa.* Retrieved from http://www.gg.rhul.ac.uk/ict4d/elareport.pdf

Urey, G. (1995). Global financial integration: The role of the World Bank. In B. Mody, J. M. Bauer, & J. D. Straubhaar (Eds.), *Telecommunications politics: Ownership and control of the information highway in developing countries* (pp. 113–134). Mahwah, NJ: Lawrence Erlbaum Associates.

UKeU. (2002). eUniversities signs launch course providers, news room, UKeU, UK. No longer available.

UK Press Association. (2009, August 10). *School dinners spy website launched.* Retrieved from http://latestnews.virginmedia.com/news/tech/2009/10/08/school_dinners_spy_website_launched

Van Auken, B. (2007). Social inequality in US hits new record. Retrieved from http://www.wsws.org/articles/2007/oct2007/usa-o16.shtml

Vanbuel, M. (2008, March 19). Why Africa cannot afford to miss the knowledge revolution. Retrieved from http://www.elearning-africa.com/newsportal/english/news108.php

Van Der Velden, M. (2002). Knowledge facts, knowledge fiction: The role of ICTs in knowledge management for development. *Journal of International Development, 14*(1), 25–37.

Velten, H. (2007). *Cow.* London, UK: Reaktion Books.

Voice of the Listener and Viewer. (2002). *Response of the VLV to review of Radio Spectrum management by professor Martin Cave.* Retrieved from http://www.ofcom.org.uk/static/archive/ra/spectrum-review/comments/vlv.doc

Wahl, A. (2011). *The rise and fall of the welfare state.* London, UK: Pluto.

Waitzkin, H., Jasso-Aguilar, R., & Iriart, C. (2007). Privatization of health services in less developed countries: An empirical response to the proposals of the World Bank and Wharton School. *International Journal of Health Services, 37*(2), 205–227.

Walby, S. (2009). *Globalization and inequalities: Complexity and contested modernities.* London, UK: Sage.

Walsham, G. (2010). ICTs for the broader development of India: An analysis of the literature. *Electronic Journal of Information Systems in Developing Countries, 41*(4), 1–20.

Webster, N. (2006). Welcome to iPod city. *Mirror Online.* Retrieved from http://www.mirror.co.uk/news/top-stories/2006/06/14/welcome-to-ipod-city-115875-17226460

Wellenius, B., & Neto, I. (2007). *Managing the radio spectrum: Framework for reform in developing countries* (Policy Research Working Paper No. 4549). Washington, DC: Policy Division, Global Information and Communication Technologies Department, The World Bank.

Weller, M. (2004). Learning objects and the e-learning cost dilemma. *Open Learning, 19*(3), 293–302.

Wellington, J. (2005). Has ICT come of age? Recurring debates on the role of ICT in education, 1982–2004. *Research in Science & Technological Education, 23*(1), 25–39.

Whitty, G. (2000, November 21). *Privatisation and marketisation in education policy.* Paper presented at the NUT conference on Involving the Private Sector in Education: Value Added or High Risk? Retrieved from http://firgoa.usc.es/drupal/files/whitty.pdf

Wilkinson, R. (2005). *The impact of inequality.* London, UK: Routledge.

Wilkinson, R. (2008, February 7). *Economic and social rights and health—What difference does inequality make?* Paper presented at the NIHRC Health and Human Rights Conference. Retrieved from http://nihrc.org/dms/data/NIHRC/attachments/dd/files/91/Economic_and_social_rights_and_health_(R_Wilkinson).doc

Wilkinson, R., & Pickett, K. (2009). *The spirit level: Why more equal societies almost always do better.* London, UK: Allen Lane.

Wilkinson, R., & Picket, K. (2010, July 8). *Beware false rebuttals.* A response by the authors of The Spirit Level to a report by Peter Saunders (Beware False Prophets). Retrieved from http://www.equalitytrust.org.uk/saunders-response

Williams, E. E., & DeMello, M. (2007). *Why animals matter: The case for animal protection.* Amherst NY: Prometheus Books.

Williams, J. (2011, July 25). IT industry salaries rise 7.5% in 2011. Retrieved from http://www.computerweekly.com/news/2240105146/IT-industry-salaries-rise-75-in-2011

Williams, R. (1986). *Export agriculture and the crisis in Central America.* Chapel Hill: University of North Carolina Press.

Wilson, G. (2011, September 6). 'Feral underclass' to blame for the riots. Retrieved from http://www.thesun.co.uk/sol/homepage/news/3797021/Feral-underclass-to-blame-for-riots.html

Winner, L. (1986). Do artefacts have politics? In L. Winner (Ed.), *The whale and the reactor: A search for limits in an age of high technology* (pp. 19–39). Chicago, IL: University of Chicago Press.

Winner, L. (1993). Social constructivism: Opening the black box and finding it empty. *Science as Culture, 3*(3), 427–452.

Winograd, T., & Flores, T. (1986). *Understanding computers and cognition.* Norwood NJ: Ablex.

Wolff, R. (2009). *Capitalism Hits the Fan—a lecture.* Northampton MA: Media Education Foundation. Retrieved from http://www.mediaed.org/assets/products/139/transcript_139.pdf

Woods, H. R. (2007, June/July). Talk and chalk. *Symmetry, 4*(5). Retrieved from http://www.symmetrymagazine.org/cms/?pid=1000497

World Bank (1991) *World Development report: The challenge of Development,* available at http://wdronline.worldbank.org/worldbank/a/c.html/world_development_report_1991/chapter_7_rethinking_state

World Bank. (1994). *World development report: Infrastructure development.* New York, NY: Oxford University Press.

World Bank. (1998). *World development report: Knowledge for development.* New York, NY: Oxford University Press.

World Bank. (2001). *Livestock development: Implications for rural poverty, the environment and global food security.* Washington, DC: The World Bank.

World Bank. (2002). *Information and communication technologies: A World Bank group strategy.* Washington, DC: The World Bank.

World Bank. (2006, September). *Word development indicators.* Washington, DC: The World Bank.

World Bank (2011), How we classify countries, available at http://data.worldbank.org/about/country-classifications

World Gold Council. (2011). *Gold demand trends Q2.* London, UK: World Gold Council.

Worldwatch Institute. (2004). *State of the world report: Consumer society.* Washington, DC: Author.

Woutat, D. (1985, June 15). Apple bites the bullet, will lay off 1,200 and close 3 of its 6 plants. *Los Angeles Times.* Retrieved from http://articles.latimes.com/1985-06-15/business/fi-12519_1

Wu, T. (2010). *The master switch: The rise and fall of information empires.* London, UK: Atlantic Books.

Wyatt, S. (2008, March 28). *Challenging the digital imperative.* Inaugural lecture presented upon the acceptance of the Royal Netherlands Academy of Arts and Sciences (KNAW) Extraordinary Chair in Digital Cultures in Development, Maastricht University. Retrieved from http://www.virtualknowledgestudio.nl/staff/sally-wyatt/inaugural-lecture-28032008.pdf

Wyatt, S., Henwood, F., Miller, N., & Senker, P. (2000). *Technology and in/equality: Questioning the information society.* London, UK: Routledge.

Yoder, S. K. (1990, June 4). Workplace (a special report): Higher tech—Putting it all together: Computer-integrated factories are said to be the savior of industry; but can anyone make the system work? *The Wall Street Journal* (Eastern edition).

Zeschky, M., Eidenmayer, B., & Gassman, O. (2011). Frugal innovation in emerging markets. *Research-Technology Management, 54*(4), 38–45.

About the Contributors

Charlotte Chadderton is a postdoctoral research fellow in education at the University of East London. Her research interests include critical race theory and whiteness in education, poststructural identities and the work of Judith Butler and social control and surveillance in schooling. She is currently writing on issues of education, race and social control.

Erika Cudworth (Ed.) is a reader in politics and sociology at the University of East London. Her research interests include the material cultures and political economy of food production, the gendering and naturing of space and place and human relations with nonhuman animals, particularly theoretical and political challenges to exclusive humanism. She is the author of *Environment and Society* (Routledge, 2003), *Developing Ecofeminist Theory: The Complexity of Difference* (Palgrave, 2005), *Social Lives with Other Animals: Tales of Sex, Death and Love* (Palgrave, 2011) and coauthor of *The Modern State: Theories and Ideologies* (Edinburgh University Press, 2007) and *Posthuman International Relations: Complexity, Ecologism and International Politics* (Zed, 2011). She contributed a chapter to *The Myths of Technology: Innovation and Inequality* (Peter Lang, 2009).

Alvaro de Miranda is a visiting research fellow of the London East Research Institute, University of East London, and a member of its management board. He was head of the Department of Innovation Studies at the University of East London (1992–1999) and a past president of the European Inter-University Association for Society, Science and Technology (ESST; 1994–1997). His recent publications include "Creative East London in Historical Perspective" in *Rising East Online*

No. 7 (2007), "Technological Determinism and Ideology: Questioning the Information Society and the Digital Divide" in *The Myths of Technology* (Peter Lang, 2009) and "The Real Economy and the Regeneration of East London" in *London After the Recession: Fictitious Capital?* (Ashgate, 2012).

Allyson Malatesta is a senior lecturer in innovation studies in the School of Law and Social Sciences at the University of East London. Her research interests include the complex relationships that exist among science, technology, innovation and society, with a focus on information and communication technologies (ICTs) and also the current concern regarding the commodification, production and consumption of education. She is a member of the Society, Technology and Inequality Research Group, the Education: Policy, Pedagogy and Practice Research Group and a fellow of the Higher Education Academy.

Miriam Mukasa is a senior lecturer in innovation studies at the University of East London, a leader of ICTs in international development programmes and a fellow of the UK Higher Education Academy. Her research interests include ICTs for development, design and implementation of information systems and gender and technology. She contributed to a chapter in *The Gender Politics of ICTs* (Middlesex University Press, 2005), which raised issues of inequality in positions by women in the computing and ICT industries, and to "The Social Relations of Large Scale Software System Implementation" in *Information, Communication & Ethics in Society* (Troubador Publishing, 2005).

Maxine Newlands is a Lecturer in Journalism at James Cook University, Queensland, Australia. Her research interest centres on discourses of governmentality, radical environmental activists movements, environmentalism, environmental governance, the uses and abuses of traditional and new media and new social movements. She contributed chapters to *Environmental Activism, Citizenship and the Media* (Intellect, 2012), *Global Environmental Politics and the Media* (Peter Lang, 2012) and *London After Recession: Fictitious Capital?* (Ashgate, 2012).

Peter Senker (Ed.) was an editor of *The Myths of Technology: Innovation and Inequality* (Peter Lang, 2009), in which he was a joint author of a chapter, and of *Technology and In/Equality: Questioning the Information Society* (Routledge, 2000), to which he contributed a chapter. He has published widely, including (with Rodrigo Arocena) "Technology, Inequality and Underdevelopment: The Case of Latin America" in *Science, Technology and Human Values* (2003) and "Technological Change and the Future of Work: An Approach to an Analysis" in *Futures* (1992).

Richard Sharpe is a senior lecturer at the University of East London in the School of Arts and Digital Industries. He is a senior associate of the London East Research Institute at the University of East London. He is a director of ETC, the UK's training provider for professionals working in media, marketing services, public relations and corporate communications. His recent publications include two chapters in *The Myths of Technology: Innovation and Inequality* (Peter Lang, 2009). His forthcoming publications include two chapters on London as a centre for financial mediation in *London After Recession: A Fictitious Capital?* (Ashgate, 2012).

Kathy Walker (Ed.) is a senior lecturer in media and communications at the University of East London, a leader of the communication studies degree programme and director of the Society, Technology and Inequality Research Group (STIR). Her research interests include the innovation and development of new communication technologies, media policy and regulation and public service broadcasting in the UK. She contributed a chapter to *Technology and In/Equality: Questioning the Information Society* (Routledge, 2000), which addressed the implications of new technologies and the transactional television environment for audiences. She was also an editor of *The Myths of Technology: Innovation and Inequality* (Peter Lang, 2009), to which she contributed a chapter.

Index

General Editor: Steve Jones

Digital Formations is the best source for critical, well-written books about digital technologies and modern life. Books in the series break new ground by emphasizing multiple methodological and theoretical approaches to deeply probe the formation and reformation of lived experience as it is refracted through digital interaction. Each volume in **Digital Formations** pushes forward our understanding of the intersections, and corresponding implications, between digital technologies and everyday life. The series examines broad issues in realms such as digital culture, electronic commerce, law, politics and governance, gender, the Internet, race, art, health and medicine, and education. The series emphasizes critical studies in the context of emergent and existing digital technologies.

Other recent titles include:

Felicia Wu Song
Virtual Communities: Bowling Alone, Online Together

Edited by Sharon Kleinman
The Culture of Efficiency: Technology in Everyday Life

Edward Lee Lamoureux, Steven L. Baron, & Claire Stewart
Intellectual Property Law and Interactive Media: Free for a Fee

Edited by Adrienne Russell & Nabil Echchaibi
International Blogging: Identity, Politics and Networked Publics

Edited by Don Heider
Living Virtually: Researching New Worlds

Edited by Judith Burnett, Peter Senker & Kathy Walker
The Myths of Technology: Innovation and Inequality

Edited by Knut Lundby
Digital Storytelling, Mediatized Stories: Self-representations in New Media

Theresa M. Senft
Camgirls: Celebrity and Community in the Age of Social Networks

Edited by Chris Paterson & David Domingo
Making Online News: The Ethnography of New Media Production

To order other books in this series please contact our Customer Service Department:

(800) 770-LANG (within the US)

(212) 647-7706 (outside the US)

(212) 647-7707 FAX

To find out more about the series or browse a full list of titles, please visit our website:

WWW.PETERLANG.COM